Making Waves

✦

Radio on the Verge

Mark Ramsey

iUniverse, Inc.
New York Bloomington

Making Waves

Radio on the Verge

iUniverse books may be ordered through booksellers or by contacting:

iUniverse
1663 Liberty Drive
Bloomington, IN 47403
www.iuniverse.com
1-800-Authors (1-800-288-4677)

ISBN: 978-1-4401-0257-8 (pbk)
ISBN: 978-1-4401-0258-5 (ebk)

Printed in the United States of America

To my wife, Jeanne.
She makes my heart sing and my eyes smile.
She makes everything better
and a better everything.

Table of Contents

Preface

Most of the content in this book appeared originally on the hear2.0 blog. For more smart analysis of radio, its future, and the future of its competitors, visit hear2.0:

http://www.hear2.com

Mark Ramsey provides an incisive interpretation of a media brand's place in its market and its opportunities to bust out of the pack and shine, along with all the audience research and brand development you need to get there.

Ramsey also consults media brands on navigating the challenges and opportunities of these fast-changing times.

Clients have included Clear Channel Communications, CBS Radio, Greater Media, Bonneville Broadcasting, Sirius/XM Radio, and major media players like EA Sports and Apple.

If you want to put the kind of thinking you read on every page of this book to work for your brand, whether it's in radio, TV, or new media, please contact:

Mark Ramsey
+1-858-485-6372
http://www.radiointelligence.com
http://www.hear2.com
or email ramseymark@earthlink.net

Foreword

Mark Ramsey's new book, *Making Waves: Radio on the Verge*, is a cutting-edge, no-holds-barred approach to radio in today's marketplace. His straight-to-the-point, one-on-one dialogue with some of the top marketing professionals in the country provides readers with important, relevant insight and viable solutions to consider as we continue to move our industry into the future.

His latest endeavor challenges the reader to redefine the way radio does business and how we, as an industry, continue to reinvent ourselves both from a digital and accountability perspective. From finding and nurturing talent and employees to redefining and reinvesting in our brand, Mark's perspective shines a much needed light on the many challenges and opportunities facing radio today. This book is a breath of fresh air in an industry that has so many exciting opportunities at its fingertips.

Making Waves: Radio on the Verge is a must-read for anyone who works in the radio industry.

Peter H. Smyth
President & CEO
Greater Media, Inc.

Introduction

"Who the hell wants to hear actors talk?"

So said Warner Bros. chief H.M. Warner back in 1927.

In one shining moment of shortsighted ignorance, Mr. Warner proved what too many of us in radio forget today: Everything is obvious, but that doesn't make it true. And even if it was true yesterday doesn't mean it will be true tomorrow.

Everything we take for granted, all the rules we live our professional lives by, can change in a blink – forever.

Radio is "on the verge," all right, but on the verge of what?

Are we on the cusp of a new renaissance, a time of unprecedented excitement and opportunity?

Or are we headed, as some naysayers argue, towards an industry-wide twilight?

I think it's the former, not the latter.

But we're not going to coast to that renaissance. We can't budget-cut our way to opportunity. The rules of our game are changing, and we had best change with them.

The fact is this: Something's happening out there.

Something big.

Obviously, we will emerge from the current economic recession intact. The advertising business – in one form or another – will revive. But that doesn't mean the pendulum is swinging back to the delirious, delightful '90s for the radio industry. Some of the dollars zooming out of radio and gently sprinkling on digital media like so much confetti are never coming back.

Never, not ever.

But radio is in the digital business, too. And the very definition of radio is fast-changing as the media world around us transforms. The question is: Can you see the changes and put yourself and your company on the right side of the tidal wave of good fortune heading your way?

Or are you assuming this is only a phase, and this, too, shall pass?

Let's begin with a harsh slap to the face, shall we?

Picture, if you will, this scenario:

1. The Economy Goes South

At this writing, oil prices are near record levels, the mortgage crisis continues to haunt us, the holiday shopping season will be lackluster, consumer confidence approaches a new low, and the stock market has the shakes and shivers.

In times like these, consumers tend not to part with their money, assuming they have money to part with. This means they're not buying the categories, such as automotive, which are key to radio's profit equation.

2. More Choices Mean Lower Ratings

Take a look at the television industry.

Recently, the writers' union returned to work after a prolonged strike. The last time the writers stopped working there were only 17 ad-supported cable channels. Today, says the *Los Angeles Times*, there are more than 100. And that doesn't count the ever-present distractions of video games, DVDs, the Internet, and DVRs.

Says the *Times:*

> *Ratings have fallen precipitously. In 1988, the top series –The Cosby Show* on NBC and *Roseanne* on ABC – were seen by about 25% of the U.S. homes with TVs. These days, TV's biggest

> shows, *CSI: Crime Scene Investigation* on CBS and *Grey's Anatomy on ABC, are on in about 13% of households.*
>
> *The 1988 writers' strike created much of the sampling that led to the explosion of basic cable and the fledgling FOX network. Today, there are many more destinations for an already splintered audience to splinter to further. And splinter they shall.*

An expanded number of options mean that the ratings of any individual option are guaranteed to shrink. We used to proclaim with disdain that eventually all radio stations would be 4.0-shares. Those were the good old days. Today many leading stations are well below a 4.0-share and in some cases the top-ranked stations are separated from the 10th-ranked station by a tiny fraction of a share point.

Let me ask you something: What's the value of "buying the top five ranked stations" when there's virtually no ratings difference between the top five and the five below that? In our world of expanding choice, our system is broken.

The greater the supply of modestly rated radio stations or substitutes, the lower the ad revenue for any such station.

3. The Taste of Accountability

Traditionally (and primarily to its agency accounts), radio has been about selling ears, not selling results. It doesn't have to be this way, of course, but it generally has been.

You know and I know that radio works. The trick is to prove it.

Thanks to Google and others, cost-per-click, cost-per-conversion, cost-per-sale, and other metrics provide an amazing amount of accountability – proof – that any given advertising expenditure creates a return on that investment. *You know what works and what doesn't because you can measure it.*

However, when a radio station sells the number of ears we reach or where we fall on a list of stations ranked according to audience size, what we're selling is *impressions,* or a *chance* for results, not *proof* of results. This is a key difference.

Increasingly, the difference between "impressions" and "results" is the difference between a lottery and a sure thing.

Online advertising expenditures will grow – from less than $17 billion in 2006, according to eMarketer, to more than $40 billion in 2011.

And what's driving that growth is *accountability*. The more marketers taste accountability, the more they'll like the flavor.

4. Arbitron and PPM – Better Precision at a Cost

Plenty of research and field experience with PPM has already shown that music stations which clean up their act between songs tend to out-perform those which are cluttered. And the easiest way to get jocks to chit-chat less about nonsense is to keep them from talking at all – or rid the station of them entirely as if they were a ratings cancer.

This, of course, is an easy, short-term solution to a ratings problem. In the short-run you get a leg up on your competitor. But in the long run you strip from the station anything that differentiates it from the lowest-common-denominator music box it has now become. And that means you make your station incredibly vulnerable to the next music-only source off the assembly line – whether it comes from the radio or online or wherever.

So clutter goes away – and with clutter goes expense and differentiation. But with clutter also goes inventory and sponsorship opportunities. Thus overall revenue may go down even as ratings go up.

Sure, you can move much of this to the station's website. But at what point does your website become a Moroccan Bazaar – a dumping ground for crap rather than a destination worth visiting?

And what of those ratings? PPM shows that station cumes skyrocket – and AQH rating takes a tumble. Now how will advertisers behave? Will they pay a premium for higher cumes – more reach – when the trend is *away* from reach and *towards analytics and accountability and effectiveness?*

Reality check from a survey by the IBM Institute for Business Value:

> Advertisers are demanding more individual-specific and involvement-based measurements, putting pressure on the traditional mass-market model. Two-thirds of the advertising executives IBM polled expect 20% of advertising revenue to shift from impression-based to impact-based formats within three years.

And what of those stations and groups that don't like the message PPM communicates? What does it signal to agencies when broadcasters can't agree on a measurement methodology? How comforting is it to spend ad dollars when stations are bashing not just each other, but the measurement tool itself?

5. Radio's New Competitors Jockey for Position

Let's pretend the industry gets its wish and HD radio takes off.

So now every station becomes three stations. And all three of those stations are rated. So now we have 100 stations "above the line" competing for the same audience that 30 stations used to compete for. Now we're not worrying about 4.0-shares, we're too busy hoping for a 0.5.

And what is the value of being top five ranked when there are a hundred stations in the ranker?

Answer: None.

Further, how likely is it that the top-ranked station in an HD world will have half a dozen HD-2 stations thrown at its format by competitors desperate to steal some of that thunder? I would estimate those odds at 100% – at least.

Let's now pretend HD dies on the vine, as I have long warned is likely. How much time and effort and expense are being invested in chasing HD as the primary driver of radio's future? What is the opportunity cost of this obsession? That is, what else could we have done with that money, that time, that hard-headed focus on our future? How diversified is your group's portfolio of futuristic ideas?

Finally, consider this: The ratings deck is stacked to favor the original players. The 100-share Arbitron world is a world which generally excludes non-commercial stations (unless you look hard) and Internet stations and Satellite Radio, not to mention video games and Internet video and all the other fabulous distractions for our eyes and ears which are carving out their own slice of the advertiser's pie today.

Your station isn't competing against all other stations. It's competing against all other entertainment distractions, some of which are likewise ad-supported. That makes them substitutes for you – your direct competitors.

Radio's 100-share pie is getting smaller, relatively speaking.

6. Disincentives to Streaming

The way the Arbitron rules are (correctly) set up, if your web stream doesn't match your station exactly, it doesn't count as *being* your station. And thanks to complications between radio, advertisers, and AFTRA, many streams are not matched to their stations.

That means every time you move a listener to a slightly dissimilar stream, your ratings in Arbitron theoretically go

DOWN. And that means our ratings system discourages us from streaming our stations at all – even as the Internet is the obvious destination for new ad dollar growth, is already part of every workplace and many homes, and will invariably make its way onto the American road at a pace that will leave HD radio in the dust.

7. The Ears of the Future

I won't bore you with the numbers because you've seen them plenty of times. But the attitudes about radio and the usage of radio among persons under 35 – and especially under 25 – are dramatically different.

Listeners who grow up with access to digital media are forever changed. They will not "grow into" us – we must "grow into" them.

The problem is this: It's easier to score among 25-54s when a format scores 35-54 because it's easier to score 35-54 than younger. But if the entire radio industry begins to chase persons over 35 – as it is doing right now – then this is a one-way journey for us all.

In a sense, radio is like smoking: If we don't create the habit when kids are young, we won't have the habit at all when they get older. And creating that habit is about more than targeting their music needs. We need to target the entire portfolio of their interests if we're to be viewed as relevant.

This is one of those realities which will sneak up on us slowly until one day when we stand back and realize that but for one or two stations in any given market, we no longer have a toe in the future.

8. It's All Just Too Expensive

In radio, when business is slow, investment is slower. Radio executives are loath to spend dollars chasing revenue gains unless those gains are not only 100% guaranteed but 100% guaranteed *right away*. That's another way of saying that investment in our products, our brands, and our futures is virtually impossible.

Next time you buy a new iPod, pay attention to the box the unit comes in. Feel the texture. Note the quality of the craftsmanship and the materials. Notice how cleverly designed the box is. As you might expect, this extra attention costs Apple money. Why spend so much effort and expense on something consumers are likely to throw away?

The answer is that the "brand" isn't simply what's in the box. The brand includes the box, too. The "brand" that consumers buy is everything surrounding the experience of that product. From the purchase experience to the box to the Apple store in which it's bought to the "Apple genius" who sold it to you to the iTunes software that powers it.

The brand is everything and everything is the brand.

And that's worth paying for and worth investing in.

Any radio industry leader who thinks that simply working harder and cutting expense is the way to weather the storm our industry is facing needs to seriously have their head examined.

We cannot cut our way to growth, only to obsolescence.

9. Where's the News?

The key to growth is to spawn and activate new ideas.

This is why the headlines go to what's cool. Because it's fresh, it's a "new idea."

New ideas require the injection of funds and talent and intelligence. Google doesn't hire the cheapest talent for the job, they hire the best. How many radio stations can say this?

So where are the new ideas in radio? What are we giving consumers to get excited about?

10. Where's the Talent?

Despite big headlines trumpeting the super-rich deals of radio talents like Rush Limbaugh and Howard Stern, you have to admit these deals are as rare as the talents who earn them.

Talent will always be scarce, of course. But is the radio industry seeking out and nurturing talent today?

Name, for example, the biggest radio talent under age 40 – besides Ryan Seacrest? You can't come up with *one?* Isn't that our fault?

Produce even one example of a radio station or group looking to the plethora of popular podcasts for a standout new talent with an already built-in base? I can't even think of *one.*

Keith and the Girl is a comedy show podcast daily and available on iTunes. While too bawdy for radio in its current form, can 1 million monthly downloads be all wrong?

Maybe this is why Internet talent tends to skip radio altogether – and go straight to TV. That's what Tila Tequila did. She was the most popular artist on MySpace – before headlining the most popular show on MTV.

The talent is all around us but we are blind to it.

So what do we do?

What follows are conversations and thought-starters on marketing best practices and trend prognostication. The emphasis throughout is on what radio can and should do, what trends are taking shape, what the consequences are, and what your response to those trends must be.

Although many of the interviews here were published – and live still – on the hear2.0 blog (http://www.hear2.com), some have never been transcribed and none of them have ever been printed – until now.

Each of these chapters is on a different theme, so don't think you need to start at the beginning. Start with what's interesting and move out from there.

Like the opportunities facing radio itself, this is a book without rules.

More Choice Can Be Deadly

***An interview with Barry Schwartz, author of* The Paradox of Choice: Why More is Less.**

Choice is good and more choice is even better, right? Isn't that what "variety" is all about? It's one of the founding principles of satellite radio and HD radio.

And it's wrong.

Some choice is better than none, says Barry Schwartz, a professor in the Psychology Department at Swarthmore College and the author of an incredible book, The Paradox of Choice: Why More is Less. *But it does not follow that more choice is better than some choice.*

What is "The Paradox of Choice"?

We tend to believe that choice is good. The more choice people have, the more freedom they have. And the more freedom they have, the better off they are. So it must be that the more choice they have, the better off they are. And this seems so obvious that it wouldn't occur to anyone to question it.

The problem is that all of this research was done in a world where choice was limited. So the question was: Are you better off with two options or one option? And the answer was always two options.

But now, we're talking about 300, 3,000, or 30,000 options. Suddenly there are so many options that people, instead of being liberated by choice, become paralyzed. They don't choose. It's too hard to know which thing to choose, or they choose badly. Or, they manage somehow to choose well, but they're convinced they could've chosen better, so they're not satisfied with what they've chosen. So that's the paradox. Choice should be liberating us, and instead it's paralyzing us.

So what is the effect of all that choice on people who are trying to make those choices?

There are three different effects.

One is paralysis, and this of course interests marketers a great deal. You go into a store. There are a million options and you end up buying none.

The second problem is you may overcome paralysis, but you end up choosing badly. Faced with an overwhelming array of options and features, consumers are forced to simplify and choose based on brand or on price.

The third problem: If you overcome paralysis and if you choose well, you bring the item home and you're convinced that you could've chosen better. You're just sure you could've done better, and the result is that you're dissatisfied even with good choices.

So paralysis, bad decisions, and dissatisfaction with good decision; these are the three different effects of overwhelming people with too many options.

You're talking about products or services for which money has to be traded for those products and services. But satellite radio has 150 choices and terrestrial radio has 30 to 40 in an average market. A mistake is cost-free and one button away from being repaired.

But you're still paying. Instead of paying with money, you're paying with time, which is even more precious than money for most of us.

So, for example, you can't even enjoy the Beatles singing *Abbey Road* because you're sure that there's another song playing somewhere out there that you're going to like even more. And when there are only five stations for you to check out, you can barely do it. But when there are 50 stations, you spend all of your time surfing and none of your time consuming.

So relative to radio, the more choices you have, the more surfing you do. Now, does more surfing mean you're more or less satisfied with what you're listening to?

I think you're going to be less satisfied because you're not really consuming. You're spending all your time trying to decide what to consume and you miss half the song.

Based on my research good enough is virtually always good enough; it almost never makes sense to look for the best. It's often a fool's errand. Realize that the only way to know that you've got the best is if you've examined all the possibilities. How do you know you're listening to the best song unless you've checked out every damn station? So it's just not possible.

How much is too much for radio? I would guess from my research that the "sweet spot" is dozens of options. That's the ballpark you want to be operating in: A two-digit ballpark.

What does this suggest for Internet radio with its dizzying variety? And for satellite radio with 150 unique channels? And for HD radio, which is all about magnifying the number of choices two- and three-fold?

I think it's a recipe for disaster for consumers unless you find a way to filter the options. Think about Amazon. Amazon is "The largest bookstore" with 2 million titles. But what makes Amazon spectacular is that it has everything for people who know what they want, and it's got this great algorithm for filtering what else it has to make sensible suggestions.

So you'd have to organize your radio options in a way that makes it possible for people to navigate so they're not looking at thousands of choices. A lot of thought has to go into how the options are organized and presented to people.

It's not that all choice is bad. It's that most of it should be hidden from you.

Lessons in Bad Branding from the Auto Industry

An interview with Charlie Hughes, co-author of **Branding Iron: Branding Lessons from the Meltdown of the U.S. Auto Industry.**

You can learn a lot about branding – good and bad – from the auto industry, and Charlie Hughes should know. He's the founder and former CEO of Range Rover of North America. He has worked for six different automakers on 11 different brands and was also CEO of Mazda.

Charlie heads the marketing consultancy Brand Rules and is the co-author of Branding Iron: Branding Lessons from the Meltdown of the U.S. Auto Industry. *Given the importance of a vital auto industry to the health of the radio industry, Charlie's got a message for us all.*

There's quite a bit in your book that's really critical of Detroit's automakers. In a nutshell, what went wrong and what can we learn from it?

It is my belief that the Detroit automakers failed to realize a rather dramatic shift in the world, and that shift is the fact that virtually every market now is over-served. There are too many car brands. The most important job is to set yourself apart.

Revenue becomes king. Unfortunately, the automakers still see themselves primarily as manufacturing companies and they are most comfortable spending time on the expense side of the equation. Of course, worrying about expenses is necessary. It's like breathing. You're not going to live if you don't breathe. But it doesn't bring you success. It's just a necessary function. So understand that setting yourself apart in this world is job number one.

Second, you need a firm grasp of who you are. It seems kind of amazing, but the industry itself hasn't agreed on the definition of "brand." I define a brand as a promise wrapped in an experience. It's based on the simple premise that in an over-branded world, before you can expand out you really have to stand for something. And that's the "promise."

The best promises are clear, compelling, and show that you are committed. They are also personal. There are promises from you as a company to individual buyers. And, of course, the companies that succeed keep those promises.

Third, the brands that succeed are those that have the most focus. Toyota is everybody's poster child for being the most outstanding automotive brand. They are swimming in cash, if you believe them, of almost $40 billion. They could buy General Motors outright. But they wake up every morning and worry about one thing – the Toyota brand and beating other people in the marketplace with Toyota. It is that kind of focus that leads to their success.

The companies that get into a lot of trouble, and General Motors and Ford are certainly two good examples, all seem to have too many brands, more brands than anybody could possibly concentrate on. And they talk with wild optimism about synergies and how we can share platforms and how the public won't notice. Well, guess what? The public does notice.

One of the key points we make is that branding is not marketing. Branding is a business strategy. The alternative to that basic strategy is to sell on price – as a generic or a commodity, and that is not a winning strategy. But if you walk down the brand strategy path, then that strategy has to permeate everything you do. It is not a badge you put on at the end of the assembly line. It's not an ad slogan. It is who you are, right down to your socks. And it has to permeate everything you do.

You say the most exciting new products get the most press and media coverage. If that's true, then why aren't there more exciting new products?

Well, I think one of the big issues in the car business is there are all these conventional wisdoms that are around, like "product is king." But if Detroit really believes that, then why do so many cars come out looking exactly the same? It is because you end up with a committee approach to building cars and people who do not really understand the marketplace and are not going to take risks. They dummy-down the things that they're doing.

Detroit and its dealerships are a huge radio advertising category. If you owned a group of radio stations right now, what would you worry about?

Because of the decline in market share of the Detroit brands, Detroit now is sitting with too many car dealers. An awful lot of money coming to radio stations comes through both dealer ad groups and individual dealers as well. So there's going to be a period where the number of domestic dealers is going to have to decline. That in and of itself may not be bad because many of the dealers who are going to be weeded out are the ones who probably don't spend a huge amount of money on advertising anyway because they're just not generating enough revenue and critical mass to do so. But that will have some relative displacing effect on local media. In that regard, I would certainly be trying to support the strong players in my market in any way I could.

Also, Detroit is trying an awful lot of nontraditional media. To the degree that they can wean themselves off heavy, heavy discounting and go to more brand-building marketing exercises, that may put a strain on radio as well. You see, radio's going to have to work to get past the image of being very good on short-term promotion and price advertising and demonstrate that it's a good tool in the arsenal of building brands.

What about what's in the dash? As you know, radio has competition nowadays from iPod jacks and satellite radio and even the coming of HD radio. Where do you see that whole thing going?

We have put customers in control and they like it. You like it in your personal life. I like it in my personal life. I have an iPod, you have an iPod, everybody has an iPod – and there are times when I want to listen to the music I want to listen to. And frankly, there are other times when I don't want to be bothered with all that and I do want to tune in to my favorite radio station.

What about HD radio? When the radio was basically the only choice in a car – AM or FM or whatever – then advancements in radio technology could play an important role. If you look at the car companies today, they've had to engineer the cars to be compatible with iPods. They've had to re-engineer their tuners so that they could either carry XM or Sirius. I'm afraid in some respects the technology has sort of diminished the importance of HD radio. Again, though, because it's a dogfight, you're fighting for attention both with consumers and also with the manufacturers. I'd be fighting like crazy to make sure that HD radio gets promoted to the car industry so that they start installing the technology. In the past, the auto industry would pick up on these things automatically because they had limited choice. Not any more.

What about the Internet in cars and its ability to deliver audio?

You're still going to see an awful lot of stuff creep in, but you may see it creep in through XM and Sirius, not through Detroit itself. The first example of this now would be the real time navigation system. The promise is that you could be driving around in your car and want a stock quote on Ford, or tell me what the score is in the Yankee-Boston Red Sox game, or I'm flying to Atlanta, is my flight on time? What's the weather in Atlanta, anyway? All that now is technically doable.

For satellite radio, most of the smart money says that the radio part of it is at most the "camel in the tent." The bigger game is in all of the possibilities for information being pumped into the car.

And when do you think we're going to see that information pumped into cars in a big way?

I don't really have a good answer for you on that. But I will tell you that it is one of those areas of life which will be insidious because once we experience it, we're going to want more of it. And that gets back to the idea of empowering customers and giving them what they want when they want it.

How to Make Radio Relevant Again

***An interview with media futurist Douglas Rushkoff, author of* Get Back in the Box: Innovation from the Inside Out.**

Douglas Rushkoff is the author of eight best selling books on new media and popular culture, including Cyberia, Media Virus, *and the latest,* Get Back in the Box: Innovation from the Inside Out. *His commentaries air on* CBS Sunday Morning *and NPR's* All Things Considered. *He lectures at New York University and created and hosted two must-see documentaries for PBS:* The Merchants of Cool *and* The Persuaders. *Rushkoff is a thought leader who has been ahead of many curves in the world of technology and pop culture, and now he turns his sights to radio's future.*

What do you mean by getting "Back in the Box"?

I get all these phone calls from business people who ask me to come and help get them out of the box. But what do you mean get "out of the box"? Well, they want me to come up with new ads or a new image or a new sense of purpose. And I ask them what they do; do they make shoes? If they make shoes why don't they just figure out how to make better shoes? Rather than trying to think of how to repackage or re-brand your shoes, why don't you make better shoes? Isn't your product itself a better communications medium for the quality of that product than some ad or some marketing scheme or some brand image?

You argue that consumers – listeners – are more than a "target."

I think what we're starting to realize is that consumers are not just targets to be manipulated but they are really members of your company's culture. They are part of sneaker culture or shoe culture or computer culture or cell phone culture or the culture of your radio station. And they're going to want to sign on to a company that can actually accept their contributions and participation in some sort of real way.

Most stations are looking at the listener community as a bunch of consumers to be segmented, targeted, manipulated – the sort of spreadsheet approach to radio as opposed to the passionate approach.

On this end of the spectrum you have a whole bunch of market research, you run your stations at the lowest possible cost, you automate all your processes, and that allows you to get the most bottom-line revenue. How do we reduce the number of offerings? How do we do high volume? This sort of scorched earth policy may generate income in the short term and will be great for shareholders this quarter. But it's a bad approach for your entire industry in the long term.

The problem with "let's build the biggest business we can, whatever it takes," is it doesn't work unless you're just acquiring things. What going public really means is you're no longer in the business that you used to be in. Now you are a brand name. You're a "word," Clear Channel or Avon or whatever, that is trying to get people to invest more money in that word. You're essentially a holding company now which is something very different. So your strategy evolves into doing the kinds of things that will get shareholders to put more money into the "word." And that's going to mean increasing certain kinds of measurable numbers in the shortest term possible. Yet that's how you can attract a real community of listeners, of participants, of fans.

On the other end of the spectrum you'll have someone saying, "You know we're an interesting group of people at this radio station; why don't we figure out what kind of music we would like to share with people, become experts in those kinds of music, and then do shows that really do explore and expose our listeners to our expertise and our passion?"

So you're saying the strategy for a radio station should be to surprise, delight, and lead, not follow?

The great surprise and bizarrely out of the box thing I'm saying is be a true expert in the thing you actually do, and then you will do that thing the best. In this new age the reward will go to people who do things the best rather than people who destroy the actual industries they're in. Getting back in the box will turn out in the end to make more money and create more success.

I think we've reached the law of diminishing returns – the Wal-Martization, the Radio Conglomerization of our world is finally failing. These companies are crumbling under their own weight. And consumers are desperate for people who actually remember the age old crafts of everything from making shoes to making radio.

How does radio magnify the impact of a culture of radio station fans? How do radio managers and programmers get "back in the box"?

If you really are an expert in radio now you don't have to be scared of your consumers any more. You don't have to be scared of your employees. How do we, as a community, build the best radio station we can? How can you honestly involve listeners in the development of your emphasis rather than just, say, Sony Music or the other labels in that development?

You should invite not your least common denominator but your highest quality listener to come in and be part of your community. And that's how you train the DJs of tomorrow. You're going to have fans that understand that you're treating them with respect, and you're trying to answer their needs in a genuine way.

Right now the radio conglomerates are not asking listeners what they need. They're asking what they can get people to listen to. And that's a very different question. If you really care about radio, if you really care about what this medium can offer the world, then you should ask yourself: "What need can I answer with this medium? What can I do today to actually make someone else's life better? What unmet needs are there, and how can radio fulfill those unmet needs for community, for civic reality, for music education, for development of new music talent, etc.?"

Rather than asking listeners which of our shows you like the best so we can do more of that for you, ask what role could radio play in your life – and that's not something that you can get from a focus group.

What is the role of "fun" in the business of radio?

Because of my book tours I've been in a lot of radio stations, and even from 1995 to 2005 the amount of change I've seen has been shocking. There used to be this kind of quality to an FM radio station – I hate to be stereotypical, but there was a certain kind of chick who would be the receptionist at an FM radio station. There was a certain kind of guy that worked in the album room organizing the albums. There was a certain kind of geek figuring out the emphasis rack.

But FM stations are not really like that anymore. They feel much more like almost any other office, and if you didn't see the control room you wouldn't know you were in a radio station at all. They don't ooze their culture anymore.

There was a smell and a quality and a texture to everything radio that I think was the fun of the industry. There was something so real about it. In the early days when I was a kid, you had Ron Lundy and Cousin Brucie – you just somehow knew those guys were there even though they were playing top 40 stuff. You knew it was a world of guys with records and personalities. And there's so little of that on the radio today.

There's almost nothing in mainstream radio that has that sense of this as a club of people in a cool place having a great time sharing some of their ecstasy with those of us driving to work or sitting in our bedrooms who wanted to have a taste of what it's like to be an adult who understands music, who reads *Rolling Stone*, who understands why we're fighting the Gulf War, or whatever it is. And I want a piece of that.

When I turn on the radio now I don't feel that these folks have a piece of anything that I can't get a piece of by going into Allstate to work in the morning. It's just another working stiff with some computer telling them what to play and when to play it and when to read the ads.

I don't trust the voice behind the music anymore because I don't know that he's really an expert or that he really cares. He's not part of a living, breathing, fertile culture whereas if I go online and look at these podcasts I know these people have done it not for the money but for the love of it. And radio is going to have to go a long way now to convince me that there's somebody there who cares about what they're doing for some reason other than the cash.

Finally, I would say the purpose of radio is to keep people company. And in order to keep people company there's got to be a human being on the other side of it. The more truly human your radio station is the better it is at keeping people company. And the more computerized and business-like it is the farther outside the box you'll find yourself.

Radio Marketing on a Shoestring

An interview with* PyroMarketing *author Greg Stielstra.

If marketing always worked, every radio station could always afford it. But the sad fact is that it doesn't, and Greg Stielstra thinks he knows why and what to do about it.

Stielstra is the author of an influential book called PyroMarketing: The Four Step Strategy to Ignite Customer Evangelists and Keep Them for Life. *Previously, Greg was the marketing director for numerous best-selling books including* The Purpose Driven Life, *the best-selling hardcover book in history.*

Greg, you say marketing is broken. What's broken about it?

Once upon a time we found a method that worked really well called "mass marketing" and then, when we weren't paying attention, the circumstances that enabled it changed and nobody bothered changing their approach to marketing. We are marketing by doing what we've always done without paying attention to how our environment has changed and the fact that the old tactics no longer work.

And what led you to the solution you call "PyroMarketing"?

I worked in the publishing industry for the last 15 years. Budgets for book marketing are very small, so we had to pay great attention to how we invested that money.

And I noticed that people would come to me with things they called "marketing" and "advertising," but if I did them often my books didn't sell. So I began to study why certain things worked and other things didn't, and the four steps of PyroMarketing began to emerge.

Conventional wisdom keeps telling us that "awareness" is all that counts, and that drives a lot of our marketing expenditures. But that's wrong. In 2003 alone more than 26,000 new food and household products were introduced: 115 new deodorants, 187 new breakfast cereals and 303 women's perfumes. It's no longer enough to be aware of a new product – you also need to know how it compares with all of your other choices before you're ready to make a decision. And often people decide not to decide.

How can stations move from "mass marketing" to what you call "PyroMarketing"?

I think that the best way to understand today's marketing process, the way messages are sent, received, acted upon, and spread, is to think of it as a fire. Now you need four things to build and sustain a fire: Fuel, heat, oxygen, and then the heat released by the combustion action itself.

In the PyroMarketing model I believe that consumers are like the fuel and there's money stored in our wallets. But there is also a very strong bond between us and our money; we won't give it up easily. Marketing provides the heat – the activation energy that can excite us about a new product or service. And if that marketing can excite us beyond our ignition point, the bond with our money breaks and we'll exchange it for the product or what I call, OH!2. The output then is the consumer's reaction expressed as word of mouth (positive or negative) that either causes your fire to grow or die.

Now a couple of really important things. First, people have widely varying ignition points. What will excite someone enough to buy a product won't be nearly enough to excite someone else to buy the same product. Second, it's not enough to have a product or service – you need a remarkable product or an exceptional service and that's what I mean by OH!2. If it deeply satisfies someone's need then they will tell other people about it, and your popularity will spread by word of mouth.

Step 1: Gather the Driest Tinder

If marketing behaves like fire then you build marketing plans the way you build a campfire. You have to gather the driest tinder. And this means focusing on the people who are most likely to buy your product, enjoy the benefit, and become enthusiastic customer evangelists for it. Whenever you can you should try to identify them by behavior and define that market as narrowly as possible.

The driest tinder for your station are those fans who are so devoted to you that they've signed up to your fan club or your regular listener program. You have to ask, where are the people most prone to gravitate to my format, where are they located, what kinds of people are they, what kinds of groups do they cluster in, and so on.

Today, it's not enough to reach lots of people if none of them care. You have to reach just the right people and that means very carefully profiling your audience. Any radio station collects a congregation of people around some topic. Often it is a music format, and what do they have in common? Well, their love for country music. Or their need for information in the case of a news/talk station.

Step 2: Touch It with a Match

By this I mean giving people an experience with your product or service. If you want folks to laugh don't tell them you're funny. Tell them a joke. And this could not be more important because of the increased choice that we all face. The quickest path to product understanding, the shortcut, is to experience it. So give people an experience with your product or service.

Here's a radio example for you: I think OnStar has done a wonderful job with their radio advertising. It's a complicated service that's hard to explain, but when you hear the tape recording of the woman who's in the ditch with her children talking to that OnStar advisor and getting help within mere moments, you immediately understand how and when you'd use it and you feel the relief that that woman feels.

Inherently, radio has a unique advantage in that it's almost always local. You have people on the ground, you're hosting events, you're in contact with your audience and you can help them experience your station or your advertiser's product in a way that many competing media can't.

You have to stop thinking of yourself as a radio station. Instead, think of yourself as the aggregator of the driest tinder. You are the leader of an affiliation network, a group of people who have voluntarily chosen to gather together around a common interest in your station's format. So who are those people and how else can you lead them? How else can you communicate with them beyond your radio signal? How often can you gather with them, and how else can you introduce them to your advertisers?

Step 3: Fan the Flames

PyroMarketing believes in fanning the flames, and this means equipping your customer evangelists to spread your marketing message throughout their social network. It's encouraging word of mouth. And when you equip them with tools to help them spread the word, you can make them much more effective because the fire, after all, is hotter than the match. So you don't keep growing your fire by throwing more advertising at it – you do it by first creating customer evangelists and then equipping them.

Just the other day Krispy Kreme ran a great promotion. It was called "Share the Love." If you went into the store and bought a dozen doughnuts they would give you a dozen Valentines. Each Valentine was also a coupon for a free doughnut. And so when I did this I came home and I gave the Valentine's Day cards to my kids who in turn gave them to their friends. And because their friends don't drive, whole mini-vans filled with families pulled up to Krispy Kreme to get free doughnuts. Now the brilliance of this is that they weren't doing anything they didn't already do. If you walk into Krispy Kreme and you wait in line they'll give you a free doughnut whether or not you have a coupon. What they had done was empowered me, a customer, to take an offer that was ordinarily only available in the store back home to my neighborhood and to make that offer to my friends who then went down to the store to collect. And so my one purchase leveraged many more.

In general I think big marketing budgets foster laziness. Small budgets foster helplessness. And no budget fosters creativity. I think the very first thing people should do is sit down and imagine that they have no marketing budget whatsoever – you come up with the better ideas when you imagine you have no resources.

Step 4: Save the Coals

If you save the coals of a dying fire, just fan it a little bit and – POOF – the fire comes back to life. Saving the coals in PyroMarketing means keeping a record of the people you encounter through your marketing so that you can return to them again quickly and affordably to tell them about new products or fan the flames.

Mass marketing finds buyers but lets them slip anonymously back into the crowd. Mass marketers have to use new matches to start the same fire over and over again. But if you save the coals and keep the record, a database of your buyers or listeners, you can build equity with your marketing.

Tattoo Your Radio Station on Listeners' Brains

An interview with "Branding Diva" Karen Post.

Karen Post is more than a branding expert, she's a "Branding Diva." She's author of the bestselling Brain Tattoos: Creating Unique Brands That Stick in Your Customers' Minds.

You describe a brand as a "brain tattoo."

Well, a brain tattoo is a metaphor that I use to help people really understand what a brand is, so that they grasp it is not just the logo or the tagline. It is the absolute sum of everything an organization or an individual or a company does that touches the marketplace. When you think of a tattoo on a body part, it's put there by choice. Brands are put in your mind by choice, because you either relate to them, they express who you are, or you look at them as a friend. The brain tattoo is a metaphor – the emotional mark that lodges in your mind about someone who tries to get you to connect to what they are selling, whether it's an idea or a product or a service.

What are your four main components of branding?

First, you must clearly understand why you are here. Is it to make money? Is it to deliver information to a community? Is it to educate people? Why are you here?

Second, you have to list your points of difference. I suggest that you create your own category. To a radio station, you come up with a new name, a new metaphor for what that radio station is so that you are creating your own category. The more distinct you can be, the more effective your marketing dollars will be, and I believe the easier it will be for people to remember you.

When I say "points of difference," you need to look at the list and say, "Can my competitor copy this easily?" If the answer is yes, then you need to go deeper and try harder and really do something that would be difficult for someone to copy.

Third, the personality. I don't mean the personalities on the station, I mean the personality of the brand. If you are introducing one friend to another, you wouldn't say, "Oh, Joe, he's got two arms and two legs and brown hair." You would use adjectives to describe how he behaves. "He's a risk taker. He's got a great sense of humor. He's a little crazy." Whatever. So think about adjectives that can describe your brand, human-like adjectives, and then do things that are consistent with this personality.

Finally, the promise. That's the flip side of the purpose. Where the purpose is logical, the promise is emotional. Southwest Airlines is a great example. Their purpose is to provide transportation and shipping, but their promise is they give people the freedom to fly, the freedom to move their product around the country through their shipping.

That's where the branding begins. Then everything you do must be evaluated against that list. If it's on brand, go with it. If it's not, it's not a good way to spend your money or your time.

Radio stations generally view themselves as bundles of attributes or features. What is the difference between knowing who you are as a brand and tweaking the mix of features and elements?

I think there's a fine line with tweaking enough so that you've got enough revenue being generated based on your programming and trying to please everybody and being a big, diluted brand that really means nothing to anybody.

As business people, we try to not miss one customer. I don't know that that's really the best attitude. I think it's better to focus on a group of people that have like interests and values, then those listeners love you so much that they become ambassadors, evangelists, of your brand. Then they make your job easier.

If you're involved in an organization that really doesn't understand the power of the brand – if they're purely, purely a sales-driven machine – then it's going to be a tough situation to really implement branding strategies. That's just like oil and water.

In your book you emphasize using the five senses in a branding effort. Obviously, radio stations are heavy on the ear, but are we effectively using sound in our branding?

Probably not. I owned an ad agency for 20 years in Houston, and if I look back I can remember what radio station press kits looked like. They all pretty much looked the same. There's an opportunity to stand out and to be distinct. Even how you package your media kit. It can certainly incorporate sound, whether it's some sort of a CD that's included or some sort of mechanism that plays sound when the folder opens. Is the paper scented? Is there some sort of specialty item in there that has taste to it? I think there's a lot of opportunity there. It goes back to being creative and doing things that are different.

You must have a sense, as a listener as well as a branding and marketing professional, about some of the things you like regarding the radio station branding you hear.

I don't know if you want to ask me that question. I'll tell you some things that I despise. I cannot stand when DJs are paid to ramble on about some product. They make it sound like it's really coming from their heart, and you know it's a paid insertion. The average consumer may not get annoyed by that, because they don't understand how that works, but that annoys me.

I'm very turned off by uncreative radio spots, where people take the old car-salesman approach and they're just screaming and yelling. People don't need to scream and yell at me to get me to pay attention to their product. They need to do something that I think is cool or remarkable and is going to enhance my life.

Your book recommends that marketers invest at least 5-10% of their company's operating budget in brand building. What do you say to those companies who plead poverty?

They're just full of hot air, basically. Every organization has money to spend on things, or they couldn't even be in business.

What they need to do is look at where they are spending. I find it hard to believe that there is absolutely no money. Still, I think people have to get creative and take some initiative to infuse the brand into those things they are already spending money on.

Also, I think people use, "We don't have a budget," as a lame excuse not to come up with cool ideas. A lot of really big ideas are free or very inexpensive. I think it takes dedication and instilling a culture of creativity in the radio station. I think that's key.

How to Make Your Marketing "Stick"

***An interview with Chip Heath, co-author of* Made to Stick.**

Why is so much marketing money flushed down the toilet? And what could be done to make it work a lot better?

Chip Heath is co-author of Made To Stick, *a wonderful book that boils down the communications and messaging process to a deceptively simple and potent formula.*

Chip is also a professor of organizational behavior in the Graduate School of Business at Stanford University.

Chip, how can radio stations create ideas and messages that connect with listeners and "stick"?

Let me give you an analogy. In 1961, John F. Kennedy proposed that we as a nation put a man on the moon and return him safely to Earth within the decade. And that was an idea that really caught on and inspired thousands of people and dozens of organizations. We were pulling together as a nation for seven or eight years to make that happen. Simplicity was one of the qualities of that idea. It was also an unexpected idea – it was fantastic and sounded like science fiction at the time. It was an incredibly concrete idea, and by concrete we mean you can picture it in your mind, you can imagine what success looks like. (Too often in organizations our goals are not nearly so concrete. We talk about increasing market share or maximizing shareholder value, but people could disagree about what those things meant. Nobody could disagree about man, moon, or decade.)

It was an emotional idea – it inspired people's passions. We wanted to achieve the next frontier; we wanted to pull together as a nation. It was a credible idea because it was coming from the president of the United States, and it was a little story in miniature: We're going to adopt this challenge, and we're going to succeed at it.

Together, those six principles form the backbone of the book. We found over and over again as we studied sticky ideas that they used at least some of those six principles, and the more sticky ideas tended to use more of them.

We boiled the formula down to an acronym: S-U-C-C-E-S.

Give me an example from the marketing world.

Everybody knows Jared Fogel, the Subway sandwich spokesperson. Jared lost over 200 pounds by eating Subway sandwiches. Now, if you tick that off against the S-U-C-C-E-S framework, it's a simple idea, it's quite unexpected. I mean here's a guy dieting with fast food, and that's not typically something we associate with fast food. It's credible because the guy who used to be overweight is giving us advice; it's concrete because we can conjure up those mental images of Jared holding out those gigantic pants that the used to wear; it's emotional because it's inspiring; and it's a story – it's a little challenge plot, like Seabiscuit or David and Goliath – Jared is tackling the challenge and overcoming it.

Nobody remembers Subway's campaign before Jared – it was called "seven sandwiches under six grams of fat." Now, that's like the information a lot of us put into our PowerPoint presentations or into our appeals to listeners. The numbers behind it are factual, you know, we do have seven sandwiches under six grams of fat, but it's also pretty abstract. It fails almost every aspect of the S-U-C-C-E-S framework. No wonder the Jared campaign worked and caught on like wildfire, and why "seven under six" is something most of us can only vaguely remember, if at all.

In general, we consumers are not good at numbers. So when we're talking about numbers in a marketing context we often have no clue what those numbers mean. That's why we must translate them into very concrete terms. For example, a medium-sized buttered popcorn contains 37 grams of saturated fat, and a group of nutritionists was trying to get that number to stick with the American people. What they did was brilliant. They made it very concrete and unexpected, and tangible. They said that medium-sized buttered popcorn has the same saturated fat as a bacon and egg breakfast, a burger and fries for lunch, and a steak and potato dinner with all the trimmings – combined!

So if I say there are a few million HD radios in the hands of consumers, that's fairly fuzzy. But if I say the average consumer is more likely to die in a motor vehicle accident than to own an HD radio, this concrete fact becomes unforgettable?

Exactly. We're going to be better off getting facts and numbers across by making them tangible, by making them human scale, by translating them into everyday terms that all of us can relate to.

What do you think is the biggest mistake radio broadcasters make in their marketing to listeners?

The message is too superficial. Don't just tell people what you do, but help them understand the identity they have that should make them care about what it is that you do.

Here's an example: The State of Texas Department of Transportation wanted to get people to litter less. And lots of research had found their target litter-bug was an 18- to 30-year-old truck-driving male. They called that "customer," affectionately, Bubba.

But how do you tap into Bubba and get him to care about roadside litter? As it happened, they coined one of the most famous taglines in Texas advertising history: Don't Mess with Texas. How do you get an 18- to 30-year-old truck-driving guy to care about litter? You make it a matter of patriotism. And Texans, above all, are very patriotic: I was born in Texas, and we believe in the United States of America and even more we believe in Texas.

With "Don't mess with Texas," you've told "Bubba" that "litter is unpatriotic."

So what does "40 uninterrupted minutes of music" mean to you as a listener? Maybe it means you're a pretty sophisticated music lover, you're not content with a bunch of talk and a bunch of DJs there yapping at you, you love music. And so, instead of just telling me "40 minutes of music," help me get a sense of what that says about myself. Texas solved that problem with "Don't Mess with Texas," and the litter on Texas roadways dropped 81% in five years. That's a pretty remarkable achievement for a public service campaign.

Your radio station must form an association between what kinds of songs you're playing and the identity your listeners want to have of themselves.

You need to boil down the marketing-speak into a very concrete image of who your customer is. That mental image will help you make very clear decisions that are consistent with attracting the kind of customers you're trying to attract. Everyone in your organization will get a common mental image about who that person is like.

But radio is bought and sold largely on demographics. What's wrong with that?

As long as we're stuck with demographic information, I don't think we're going to be as creative as we could be, because we're really not feeding into our brains the kind of material that our brains work best with, which is concrete images of specific people doing specific things.

Here's a great exercise for radio station managers. Ask yourself who is your prototypical listener, the primary listener for the kind of station you're pitching?

Let's say he is a graphic design artist who works downtown, who reads the local independent paper, and tends to have more varied taste in books and music. Tell me what he reads during his lunch break, tell me where he gets his coffee. Tell me who his favorite music groups are. And so on.

Pretty soon, you'll have a very concrete image of that customer, where to find him, and how to appeal to him. And the more concrete you can make this image of your customer, the better you'll be at making key business and marketing decisions.

Radio's Marketing Sucks

***An interview with author Mark Stevens, author of the best-selling business classic* Your Marketing Sucks.**

Mark Stevens is head of MSCO, a hybrid marketing and consulting firm, and author of the best-sellers Your Marketing Sucks, Your Management Sucks, *and* God is a Salesman.

Your first book is titled *Your Marketing Sucks.* Why does so much marketing suck?

Most marketing doesn't generate return on investment, and what marketing really is supposed to do is grow a business. It's not supposed to create sexy commercials or public relations stunts or create controversy. It can do all those things, but only if the endgame sells products and services. So marketing over the years has been distorted. The best example of that occurs in what I call the "Stupid Bowl," not the Super Bowl.

So much of marketing today is only aesthetically driven: "Can I make a beautiful ad? Can I make a sexy ad? Can I show a car?" – watch TV tonight, and you'll see 15 car commercials. Each one will be the same. You don't know which car it is. This drive for aesthetics instead of return on investment has damaged what is a very important part of the business process – marketing. And whether you're a small company or a very large one you've got to engage in marketing that generates return on investment. It has to generate more than you spend. And that's been lost.

So how do radio stations create marketing that generates more than we spend?

Basically you need two things: A killer application – something really strong about the product that makes people want to buy it in a world filled with me-too alternatives, and a killer offer – one which stands out compared to competing offers.

For example, if we're selling a teapot it might be one which automatically purifies the water to make it as clear and clean as spring water (the killer application). And you're getting a year's supply of tea free (the killer offer).

And the offer should make it hard for people to resist. So you tell people that if they want to learn the secrets – here's the telephone number to call to order the teapot and/ or the website to go to. But you also tell them, "For our free booklet on 25 ways to make a pure, better cup of tea – a reinvented cup of tea – go to this number or this website, and we'll send it to you free." Why are you doing that? Because you're better able track whether or not people are listening to the commercial. Are they responding to it? You also get their email address when they ask for the booklet on the better way to make tea. You've now captured them as a prospect and can go back to them. And you'll know – right away – if you've succeeded by their response.

There's lots of talk about accountability in advertising now, but accountability isn't just about how radio is measured – how heads are counted – via diaries or meters. It's about saying "listeners bought this product because of this commercial." And that is the job of the station and the agency, not the ratings company.

Radio has to sell more than time; it has to sell ideation to help the advertiser be more effective and successful. Everybody's afraid of Google, but radio shouldn't lie down at all. There's something about radio that Google and television cannot do: Radio is very personal. You can't beat the personalization of radio.

What's the best way to use radio as a marketing medium?

When you're marketing using radio you have to always recognize that it's a personal medium. So we like our clients to either tell a story in their radio commercials, or where possible, have the personality tell the story.

We're about to launch a product on *Car Talk*, for example. We're going to have the hosts, in a live read, direct their listeners to a website. Only in radio – that is so powerful – two well-respected figures who are able to drive people to a website in a personal conversation with their listeners. Then we have to make sure we can measure whether people are coming to the website. Are they leaving their email behind? Are they asking for the coupons that we're gonna be giving on this particular campaign to buy the service at a discount? And if they're not, the silence will tell us that we're failing.

But in most agencies, nobody really cares, because "we got an award for the commercial." The agencies forgot how to sell.

But the stations say the agencies and buyers view us as a rank and only a rank. All they care about is where we're ranked.

If you simply accept the status quo, then you just say, "I'm not going to try to increase the spend on my station."

Anybody who's ever done anything great in business has stepped out of the pack and said, "I'm not gonna accept the status quo." And there's always been some risk. But today, so much of traditional advertising isn't working, and the babies are going to get thrown out with the bathwater unless people intelligently challenge the status quo, step up to the plate, and say, "Look, listen, I'd like to talk to you respectfully – it's not just rank. I can give you some important input into the mind. Let's do a little test together. Don't spend another dime with me in the beginning. I'll show you that I can add value."

People will listen if you talk to them that way. And maybe you can't talk to a 21-year-old media buyer, but you can certainly talk to a vice president of media buying. And you can call him up and say, "Can I come in and talk to you and take you out to lunch? I have an idea I want to discuss with you. Don't spend another dime with me than you're doing now. But let's try one test to spend it differently because I think I can add something to you."

How Radio Can Confront Change

***An interview with venture capitalist Pip Coburn, author of* The Change Function: Why Some Technologies Take Off and Others Crash and Burn.**

Pip Coburn studies change for a living. He's the founder of Coburn Ventures, a trend-aware media investment company, and the author of the book The Change Function: Why Some Technologies Take Off and Others Crash and Burn.

As the radio industry meets its future, it's important to know how to make the right decisions the right way. The sad fact is that the vast majority of new technology products fail. But Coburn has found the alchemy that, correctly applied, can turn all our strategies to gold.

What is the "Change Function," and what does it mean to the changes coming to radio?

Over 90% of new technology products fail, and I don't think that's necessary. There is an inherent belief – an assumption – that you create these massive technology-disruptive changes and then you wait for price to fall. Disruptive technology plus falling price – you put those two things together, the thinking goes, and magic will happen – the markets will take off.

But as a consumer, I want to know what I'm going to hand over a dollar for. Because if I don't hand over a dollar, then you have a hobby, not a business. So a marketer really needs to ask why consumers will hand over money for something they aren't doing today. Why will you, the consumer, change your habit?

Specifically, I'm interested in what your current "pain" is. On a spectrum from "indifference" at the bottom to full-out "crisis" at the top, what is your level of "pain"? Now, what is the "pain" associated with adopting the new technology solution being offered? If the pain of changing is greater than the pain you're experiencing right now, you're not going to buy that new gadget. But if your pain now is higher than the pain of switching, then you will make a change.

The important thing is to construct your marketing to help people understand that they really do want something, and then reduce the pain of adoption by helping consumers understand how easy it is to adopt the new technology product (assuming it is).

So for satellite radio, the consumer is balancing the "pain" of commercials and more limited choice on terrestrial radio, let's say, against the "pain" of adoption, which includes the hassle of installation and the monthly fee?

Generally speaking, price is a small part of the whole "pain of adoption." Most of the "pain" has to do with the process of buying the radio: Waiting in line, installation, the learning curve, the anticipated need for help. These are "pains" – literally.

And your local radio stations have services and functions that satellite can't match.

The question is this: What can local radio stations deliver in their advertised world that the user still receives from them uniquely?

You can drill down and realize what people are really using your service for. Then you start getting the marketing message clear on what that thing is. You figure out what the differentiator is that satellite radio can't provide, and your marketing would gear towards that.

I have both satellite and traditional radio in my car, and one of the things you don't get from satellite radio that you do get from local radio is that sense of community.

For example, local DJs can create a common bond with listeners whereas if my satellite radio is just playing song after song of country music, there's no sense of community. The iPod is great, but it's not the only answer for music fans – there is value in having someone guide you through that music. And this is what you lose when broadcasters try too hard to cut costs.

What do you think about the prospects for HD Radio?

HD Radio seems like a lot of effort for the user. They're not picking up the HD radio when they buy a car, for starters. So, it's an extra trip. The up-front price is about the same as satellite. And HD radio breaks up the band by layering additional stations along the same spot. With satellite radio it's simple: 1 through 120. But for HD radio I have to understand a new world where there are multiple stations in the same spot where there used to be just one.

The other issue is this: Can the local radio stations in any given market work together so that my dial doesn't become like a bag of melted caramels? Satellite radio is organized: The rock stations are next to each other. The decades are next to each other. You get all the baseball games in one area of the dial. Can HD radio do this? If not, I think it's going to be very difficult for consumers. It feels like it's going to be messy.

The HD radio message isn't really clear and simple, and it isn't perceived as being easy. Maybe it's also a function of marketing dollars. In the satellite radio case you have two companies, and satellite radio is all they do for a living. They sent these satellites up into space. They have a lot of money to put towards marketing. They have time. They have some patience. And HD radio doesn't seem to have any of these elements.

What is your distribution? What is your name recognition? What are people talking about? Which of those two products, satellite or HD, is closer to having some peer pressure associated with it? Satellite has a huge advantage. It already has 20 million units installed.

So how do you react when someone claims that when the price of HD radio receivers drops to $100, sales are going to take off?

It's rarely, if ever, about the price. It's not that price doesn't matter at all, but if people aren't interested in buying the thing you're selling, bringing the price down on a new technology will not solve your problem. Helping them see the value that they're getting for the money is far more important, and so is reducing their perceived pain of adoption.

So what does HD radio need to do?

First, they need to differentiate the service from satellite: What makes HD radio different enough that you're willing to tolerate a few commercials, you're willing to do without the Comedy Channel and Howard Stern and Major League Sports? What's the side-by-side comparison?

Second, they need to fundamentally understand why people use radio and what it is that we can provide that is really the heart of what people want and isn't redundant to satellite. What crises do potential radio consumers have? What are the ways to sell these people? What are the services they want, deep down? The sense of community is absolutely one. People do not like to be isolated, and radio has been bringing people together for 100 years now. Safety, family, health; there are a lot of issues that radio provides to people and that satellite radio isn't doing a good job of right now.

Finally, they need to connect very, very quickly with the car companies so that when you go into the showroom you get your choice between two things competing side-by-side: Satellite vs. HD.

Here's the reality about satellite radio: Pretty soon every car will feature the choice of satellite radio or not. And there are 17 million people a year walking into car showrooms. For them, satellite radio will become an impulse purchase, and the price becomes irrelevant – it becomes virtually free. It's fantastic to be selling in that particular environment because it's painless to the consumer. It's not an extra trip.

Radio Advertising: What Works and What Doesn't

***An interview with Rex Briggs, co-author of* What Sticks? Why Advertising Fails and How to Guarantee Yours Succeeds.**

Yeah, you can wait to see how your ratings respond, but is there any better way to know what advertising works, "what sticks," and know before you spend the big bucks? And is there any better way to get improved results for your clients – and more of their advertising bucks in the bargain?

Well, yes there is. And Rex Briggs has written a book on the subject. Rex is the co-author of What Sticks? Why Advertising Fails and How to Guarantee Yours Succeeds *and the founder of Marketing Evolution, a marketing effectiveness, research, and consulting firm.*

Rex, when it comes to radio advertising, what does stick?

In radio, what's been really successful are the ads that are able to paint a picture, even through the words, and are able to connect the message to the main part of the campaign running in television, magazines, or other media. When they reinforce each other it works wonderfully.

What often doesn't stick in radio is a completely different stand-alone ad that doesn't connect back to the imagery or the sound imprint that's coming from other parts of the message in other media. That can often create a discontinuity.

So what you're saying, essentially, is that for those radio spots which have TV equivalents, it's best to use audio elements from the TV spot in the radio spot in order to connect back to the original campaign.

If a consumer hears a message on radio, and they see it in a magazine or on television or on the Internet, will they know that this is the same message? Will it connect and fit together?

That's what we call "surround sound" marketing, which is when each of the different marketing elements reinforce one another to create a more powerful experience for the consumer sitting in the middle of all of that advertising and messaging.

In radio we tend to believe we should put all our marketing money in one media basket to "own" TV or outdoor or direct marketing. But you indicate that it is much more effective to allocate resources across a mix of advertising vehicles, not to try to own one outright.

That was one of the shocking things we discovered. You see, consumers are pattern-recognizing beings. Our brains are naturally looking to see the same message or the same idea in different environments, and when you see that reinforcement of the same idea in different places, it actually creates a deeper meaning in our minds than if we just see the same message repeated again and again and again in the same environment.

So you are better off recognizing that there's power in the media mix, and that power is greater than if you concentrated only through a single megaphone. Think of the difference between mono and stereo, where the sound is coming from all different angles. Who wants to go back to mono, right? That's like going back to a TV-only plan.

What else "sticks" for radio advertising?

The most important issue is for the marketer to stop for a moment and decide what exactly "success" is. What's our business success and what's advertising success, because those aren't always the same objective.

For example, in radio, let's say that you're part of a campaign that's meant to drive foot traffic to a retail store. Well, the business success for the marketer probably isn't foot traffic, it's probably sales or profit, and having the marketer recognize the difference between the business success and the advertising success will help them recognize how to properly measure if the advertising did its job.

Now let's say the foot traffic isn't there. Then we need to think about the creative, because it's not necessarily that radio is ineffective in driving foot traffic, it probably is the fault of an ineffective message. And that's where advertising pre-testing comes in. We would try different radio ads in different markets and see if some messages work better than others.

If you're one station in one market you might try running different ads in different weeks, and correlating the spots run to foot traffic. It then becomes collaboration between the station and the marketer, the local merchant, or agency. You're out to achieve your client's goals rather than sell them a schedule of advertising per se.

A healthy collaboration with the clients will achieve better results for them and for you. We believe in the effectiveness of radio and we know the medium works. The question is, what message do you put inside the medium, and how do we make that as powerful as possible?

On radio you said the marketer has to figure out how to convey meaning with sound. What do you mean by that?

As consumers we have become increasingly visually oriented in the way that we deal with communication, and visual symbols play a very large role in the way that we communicate in advertising and marketing now. And that has led to some bias in media plans towards the visual communication media: Television, magazine, Internet, and even more so towards media that have a motion component to the visual. So it has been a challenge to show marketers that sound is actually an opportunity to differentiate the brand and reinforce meaning. Some companies have done a really good job of coming up with a sound imprint. Look at Intel as an example. Their sound imprint helps to reinforce other areas where you've gotten the message about what Intel stands for.

Or think about Wells Fargo. I remember a radio ad that used the same sound of the cowboy, or whatever he is, who's riding the stagecoach saying "Yah" to the horses, and you hear the horses galloping and moving on, and that reinforces the idea of "fast then, fast now," which was their message at the time. And that worked very well in radio because that sound imprint tied back to their overall message of "fast then, fast now."

Unfortunately, a lot of marketers don't spend nearly enough time thinking about all five senses that you can communicate a brand with, and sound is one of those that is under-leveraged.

What is the main opportunity for radio in communicating its value to potential advertisers?

My sense is that the big-spending national marketers of brands that under-leverage radio haven't really thought through what their visual equivalent is on the radio – the audio imprint of their brand. A lot of them have a visual identity, but they haven't developed an audio identity. So, I might really try to figure out how I became the leader in helping agencies and marketers determine what that audio imprint is.

Out my window here in our California office I see a UPS truck driving by, and I can tell it's UPS even though it's pretty far away because it's a big brown truck with a yellow logo on the side. It has a visual identity that you can see from a distance, but what's the audio equivalent that says UPS automatically to me? If I could help UPS develop that, and show them that it worked to reinforce the key promise of their brand and help them drive their business success, I could certainly get more of their advertising dollars.

What do you say to the radio station that says, "Look, all I need to do is show my spot in the ranker," and to the agency that says, "Look, radio station, all I want to know is where you are in the ranker."

I would say to look at what happened with vendor and supplier relationships as supply chain theory kicked in. And what happened was that more and more manufacturers whittled down their lists of suppliers to the ones that were really adding value beyond just delivering the products to spec, because that is the exact same thing that was happening in that industry 10, 15 years ago where they thought, "All we have to do is deliver the raw material to the spec we promised and show that we deliver the quality that they specified."

And a lot of those companies aren't in business anymore.

Create an Accidental Radio Brand

***An interview with David Vinjamuri, author of* Accidental Branding: How Ordinary People Build Extraordinary Brands.**

David Vinjamuri teaches at New York University and is president of Third Way Brand Trainers, a marketing training company whose clients include American Express, Starwood Hotels and other leading consumer brands. David formerly was a brand manager at Johnson & Johnson and Coca-Cola.

David, what is an "accidental brand," and if it's accidental, how do you create one?

An accidental brand is one that's started by somebody who doesn't have a background in marketing and who's solving their own problem. I call them "accidental brands" because it's usually some kind of a fortuitous accident that gives you the idea for the problem that you can solve. A good example of that is Gary Erikson, the founder of Cliff Bar. Gary was a very avid cyclist, and he was on a 120-mile ride, which seems crazy to me, but apparently for elite cyclists, that's not unusual. Now the physiology for elite cyclists requires them to keep eating while they're biking to replenish their body's store of energy.

Gary had eaten five Power Bars already on this day, and he went to eat a sixth when he realized they had another

60 miles to go and he wasn't going make it without more energy. He told me he actually started to wretch. He physically couldn't swallow it down.

And it's nothing against the Power Bar, because the metaphor for Power Bar was "fuel." But Gary thought, "Wouldn't it be nicer if the metaphor was 'cookie,' and this tasted like actual food?" And he realized he was in the unique place to solve this problem because he knew a lot about bicyclists, he knew about manufacturing, and he also had a little entrepreneurial business that had never made any money on the side that was supplying baked goods to San Francisco area bakeries.

And so he created the Cliff Bar.

Well, it's interesting that you say entrepreneurs like Gary set out to solve a problem that they themselves had – that they essentially *were* the consumer. It wasn't that they were targeting some group of people. They *were* the consumer. This is interesting first because it illustrates the importance of having a brand that actually solves a problem.

Yes, absolutely.

And second it seems to me that the entrepreneurs who solve these problems are passionately devoted to that solution.

Yes, I feel from having talked to many of these people that the passion comes the more they investigate the solution and the more they experience the problem. I think it's important that they actually experience the problem they're solving and it's for the two reasons that you said. One is because it gives them the right instincts as a consumer. There's a lot of controversy over this issue of following your gut instinct. You may have read Malcolm Gladwell's book, *Blink*, and he's been criticized because a lot of people rightly see in their life that there's been a number of times where they've made bad decisions based on gut instinct.

What you see in accidental branding is that your instinct works when it's something you know intimately, that you're connected to. And that's why I think being an entrepreneur who's solving his own problem is important. Now the second thing you mentioned, which I completely agree with, is that it has to solve a problem because entrepreneurs are already up against a steep challenge to begin with. Not being known, not having resources – all of those things.

The additional challenge of not solving a problem puts too much of a burden on you. You need to do something that somebody hasn't done before or do it in a way that somebody else hasn't done before. Because if you don't have that insight to something that's not there and needs to be there, really – I would say go work for somebody else.

You say do something that somebody hasn't done before. That sounds like your Rule No. 2, "Pick a fight," which really means doing something different from what the other guy is doing.

When I say "pick a fight," I mean taking a step beyond that and using somebody else as a way to define what you stand for. So if you don't like the way that Starbucks treats people by their long lines or the flavor of the coffee or whatever, you say, "Well, I'm not Starbucks. *This* is who I am." And by picking a fight with them and pointing out your differences you're defining your own values.

Is this something that would be literal in the branding – in the messaging?

It can be. You know, a good example of an entrepreneur who has done this very literally is Steve Jobs at Apple. He originally made IBM the enemy, and if you remember that iconic 1984 spot, it was actually focused against IBM when it looked like they were "Big Brother." And years later when Jobs came back to Apple, he picked Microsoft as a target because they represented the things he didn't agree with.

How is asking consumers what they like different from actually *being* the consumer and creating what you, yourself, like?

Sometimes when you ask people what they like you don't get the right answer. Look at Herman Miller, an office furniture maker. In the late '90s they came up with a new design for a chair, and it was radically different from anything they had created before. It used see-through fabric. It was breathable. It was more ergonomic. But it looked very, very different from anything anyone had previously done. When they initially came out with this chair, the reviews were horrible.

People thought it was ugly. It didn't make sense. It looked ungainly, and it really didn't sell.

But Herman Miller did not give up. Instead they placed the chairs into some of the right environments – places that had more modern and futuristic furniture – design firms, for example. And eventually, a culture of acceptance grew up for this product, the Aeron chair, which has been their number one selling product of all time.

So when you ask people whether they would like something and the answer was no, was it because they didn't really understand it? What if they not only saw it but actually interacted with it?

One of your rules is to be "unnaturally persistent."

Yes, one of the things people assume is that it all proceeds at a sort of a normal pace – that you sell $1,000 in year one, $2,000 in year two, and $3,000 in year three, and so on. But that's rarely the case. Most of these entrepreneurs went for five or 10 years at fairly low levels before they suddenly got a burst of awareness – they hit that "tipping point" that Malcolm Gladwell talks about – and then they exploded. But during those early years you get some progress, and then there's a dip.

And you actually have a setback for six months or a year, and you have to figure out at that point whether you're doing something that people really need and want. And if so, you have to be very persistent with it. It was interesting talking to J. Peterman, who created the J. Peterman catalog. I remember him very clearly telling me how he had been involved in this wacky venture to popularize beer cheese.

He said, "You know, I got to this point in that venture where I just didn't think it was going to work." It was a year into it, and he stepped away from it. And then when he got to the same point with the Peterman catalog, he realized he owed too many people money to get out of it. So he kept persisting at a point where he normally would have given up, and that's what made the catalog a success – that he didn't give up.

Another rule is to "build a myth."

Yes, and when I say "myth," I mean something that explains the world around you, but also imparts values – something authentic, not false. A great example of this is Roxanne Quimby, the founder of Burt's Bees. Now, there is a Burt, and Roxanne actually met him in 1984. She was living in Maine in the woods in a tent by a lake with her five-year-old twins, and things were getting a little desperate.

She had seen Burt on the side of the road selling honey from his bees in used gallon pickle jars. And he picked her up when she was hitch-hiking to go to the Post Office and they started up a conversation. She convinced him over the course of the next couple weeks to let her learn how to tend bees.

Eventually, Roxanne suggested they package the honey in cute little containers and sell it at craft fairs. And that was the beginning of Burt's Bees. But instead of talking about herself or about how natural her products were, Roxanne instinctively realized that by talking about some aspects of Burt, which were all true, she could explain to consumers in a much more powerful way what her brand was about.

And so she created this mythology of "Burt, the Beekeeper" that was about something that exists within all of us, which is the desire to return to nature, to live a simpler life, to uncomplicate our lives. And that was really what her products were, too.

So you're talking about the creation of a "story," which is bigger than the attributes or the benefits or features of the brand.

Right, it's not about features and benefits. What she was really trying to do was convey the brand values by telling a story, and the story was true.

As a radio listener, as a brand expert, and as a guy who literally wrote the book on accidental branding, how does the brand of radio or the brands of radio strike you?

I find radio fascinating. And I think that's why some of the things I learned from these entrepreneurs in *Accidental Branding* actually apply very well to your industry. The book talks about being special and unique and differentiated, and ironically the more specialized you are and the more unique you are, the more people you're going to attract.

We're always afraid of not wanting to offend anybody and not wanting to limit ourselves, and yet look at some of the most powerful personalities on radio and you actually see that they do quite the opposite.

Well, that's an interesting point, because yes, a lot of the people in radio circles are trying to hedge their bets and be as conservative as possible. But you're saying that if you take some smart risks there's more audience there for you than for someone who's taking none.

And the moment that you don't take those risks, you're probably not being genuine either, and the most important aspect of a successful brand, from my perspective, is authenticity. That's what people are always looking for.

If I'm going to buy a pair of boots, I want to buy the boots that construction workers wear that are indestructible. That was Timberland 20 years ago. I want expert brands, so when you start to say, "Well, I also need to please *these* people, and

I also shouldn't be too controversial about this," you're not really being yourself, and that's not going to help you.

So what would be your recommendation then for a radio station or the radio industry looking to bring that authenticity to the station? It's an odd question because I'm really asking you how can I be real.

No, it's a fair question because you can't do it in a vacuum. You actually have to look at the people that you're trying to reach. What do they have already? What needs are already being met? And is there something that's true about me that I can turn into a set of offerings for a radio station that would fill this unmet need?

Is there a way you can step into a vacuum somewhere or deal with something that nobody else is doing? I think that's what I would look for. And be very focused about it; that's the key thing.

In radio we focus on formats, which are one way of defining the differences between the brands , but ultimately when you're talking about authenticity, you're talking about human characteristics which may not be as narrow as any one "format," yes?

Yes, and I would argue that your same consumers are engaging in various formats anyway. They may be listening to talk radio. They may be listening to play-by-play commentary. They may be listening to music – all of those things. What you really want to do is to stand for something that they believe in and deliver something that's unique to them.

When people are really focused around that, it's very appealing, and that's also when you get publicity and attention for what you're doing.

You mean because it stands out to them, it's interesting to them, it's original to them, and it feels authentic to them.

Exactly.

Spotting Radio Industry Trends

An interview with RW Trend's Robyn Waters.

Robyn Waters is the former vice president of Trend, Design, and Product Development at Target. She led the team that transformed Target into "Tarzhay." Now head of consulting firm RW Trend, Robyn is the author of the book The Trendmaster's Guide – Get a Jump on What Your Customer Wants Next.

What are the things a radio station can do to get a jump on what their listeners will want next?

Radio stations can learn from the three elements which made Target so successful.

The first is to be "trend-right." To know what's going on out in the marketplace and know what people want or desire.

The second is to be "customer-focused," and that means knowing what your listener wants – what's going on in their hearts and minds.

The third is to be "design-driven," and that's the ultimate "secret sauce" of Target.

So the question for radio would be how you design the listening experience in a unique way to deliver your product – your story, your news, your messages – in a superior manner.

Here's what I like about radio: I love being engaged in a story, not just being delivered the facts. When you connect the facts in an engaging way, that's what pulls the listener in and brings them back again and again.

How did Target connect the trend dots using the three elements you mentioned?

By being "trend-right" they knew it wasn't just about what was next, but also about what was important.

Information is ubiquitous today. Everybody can know what's going on out in the trend universe. But what's important is what's going on inside the hearts and minds of consumers. What are their lives like? What matters to them? What makes a difference? What pushes their magic buttons?

In that way, Target wasn't trendy just to be trendy. It was translated to connect with its audience.

In radio, do you really know who you're talking to? Are you talking to an individual or a demographic or mass market?

So much of our research is focused on demographics, but customization is key. Listeners will customize their listening experiences based on the stations they listen to. But how can your station deliver some kind of a customized experience? How well can you know the desires of your listeners, not just their needs?

So much of our marketplace is focused on what people need. But the real differentiating factor is what connects to the heart and mind, what touches their life, what goes on inside, not just outside the marketplace. Products and services that live here are the kind Seth Godin calls "remarkable."

How important is it to have a passionate drive?

Passion is absolutely critical. When I was in the corporate world, passion wasn't something people talked about. I saw more emphasis placed on quantitative measures. "Let's look at the charts, let's look at the graphs."

Passion is hard to qualify, hard to measure. But boy, when it's there people know it and they resonate with it. It makes an incredible difference.

Nowadays it's not always viewed as professional to get excited about something. But at Target, if we – the product designers – can't get excited how can we possibly expect our Target guests (customers) to get excited about it? I remember sitting in a lot of meetings presenting a lot of fabulous concepts and seeing a lot of stone-faced, glassy-eyed responses. That got to be very frustrating.

So much of the "glassy stare" has been driven by Wall Street and a focus that's so short-term – next quarter, end of year. In order to reach those goals the strategy usually requires taking something out – take out that extra detail, take out the fun stuff, take out expense. That will push you towards the same products that everybody else has, the bland and unremarkable.

Yet it's exactly the extra little delightful touch that put the *Tarzhay* into Target.

That's why I always went in search of people with soul and passion. And your listeners probably do the same. I think they recognize authentic passion and I think it really pulls them in.

What do radio stations do that they should not do?

Some stations try so hard to shock or rattle cages. I think what people are really hungry here for is what one author calls "peace of mind," not more "stuff." If you can really deliver "peace of mind" and if you can do it with grace and courage, that's the kind of leadership that will help propel the radio listening experience to another realm.

Target also seems to be an entertainment experience.

There's a "softer side" to any experience. I talk about the "heart" a lot. How do you put some "heart" into a product? Some of the products that were most successful, unusual, and remarkable at Target were silly, clever, fun things that were witty and had a little soul.

There's a difference between the function of a product and the experience it allows the consumer – or listener – to have. Focus on the experience.

So in radio we should ask what we're doing to create a memorable, remarkable experience for the audience?

How can you make them feel like what you're giving them is something they personally desire? How do you customize the experience? There are lots of little ways – how you talk about something, the passion evident in your voice, etc. Listeners need to know you're talking to them personally and understand how they're feeling.

I don't think this is what people will say they want from a radio station – more peace of mind, to feel better – they will say they want more news, more information, etc., but ultimately it leads to more peace of mind.

Great example: Starbucks. You know you're going to get a great cup of coffee – you also know you're going to pay more for it. But what you're really buying is a five minute vacation. They're giving you peace of mind. They're rewarding you with what people most crave: A moment to yourself.

Given that a radio station can only speak with one voice to all listeners, how can you customize the listening experience to each person in your audience?

You have to reframe how you think about it. You can't give every person their favorite content, but if you can go back to the soul and the heart you're doing something very special for that person – you're giving them something they need. Is there some way in how you deliver the content, how you time it, the stories you tell, the content itself – is there some way to reframe it in a different way?

For example, when you give the news you could include little stories that enliven the facts and tell a tale in a way that matters directly to each listener. In other words, the facts may be the facts, but sometimes people want a story. As Margaret Mead wrote, "The human race thinks in metaphors but learns through stories." It's the stories that can help personalize the experience for the listener.

One of the popular themes in the radio business now is "Less is More." Fewer commercials, more music. Is less really more?

Even though we seem to be on this quest for more stuff – more money, more toys – when we're free of this stuff we live a better life. In the book *The Paradox of Choice,* the author discusses a market test for Smuckers Jam. Two supermarkets in two different but similar neighborhoods – in one grocery store they presented eight flavors of Smuckers and in the other, 20 flavors.

Here's what they found: More people stopped at the table with 20 flavors, but fewer people bought because they were overwhelmed by too many options. Fewer people stopped at the table with only eight flavors but more bought – the choices were more clearly defined and they could find something they liked more easily.

So sometimes more choice isn't more. It's less.

Too much "information" without editing is toxic. It makes things worse. It's what I call a "claustrophobia of abundance."

Radio is changing. HD radio is on the horizon. Will tripling the choice give people what they want or is it an illustration of the "claustrophobia of abundance"?

It's a paradox. On the one hand you get exactly what you want to listen to. On the other hand the choice can be overwhelming. As long as you know whom you're talking to, you've got a better chance of figuring it out.

How to Make Your Morning Show Funny

An interview with comedy guru Steve Kaplan.

Comedy is not pretty, and it's not easy either.

Steve Kaplan should know. He runs the entertainment industry's number one course on the topic and has been the called "The Stanislavsky of Comedy." His students have included writers from Will & Grace, Seinfeld, *movie and TV directors, studio executives, producers, and others throughout the entertainment community.*

If you want to go to school on building a great morning show, don't just attend the morning show bootcamps, go where the comedy experts go.

What can your Comedy Intensive teach an audience of radio morning people?

It breaks down what's funny and why it's funny, and we talk about how to fix it when it's not funny. We look at the "physics" of comedy, and we break it down. We find the hidden levers or "tools" of comedy, and we spend a day looking at those dynamics. Then we look at great examples of comedy and see how these tools are either used well or poorly. We're looking at writing and performance and characters, and I think that's useful for people in radio.

What's the secret to being funny?

The secret starts with making fun of yourself. You can't just make fun of somebody else. Look at Howard Stern – he's really not trying to shock you. He's just telling you what he likes, and he's telling you about himself. So he tells you how small he's endowed. He talks about his insecurities. He talks about his fears. He's making fun of himself, and by finding the humor in him you're able to laugh at him, and in a way you're able to laugh at yourself. He makes it safe to laugh at yourself.

What you're describing isn't just self-deprecating humor, it's also telling stories – personal stories.

Well, what is comedy? Comedy tells the truth, and specifically, comedy tells the truth about people. So Howard Stern tells the truth about himself and so he's telling the truth about being human. And if you look at successful comedy you'll see people who, more than anything else, have a unique viewpoint on the world but also have the ability to make themselves the butt of humor.

So if you're going to do great comedy, you're going to be telling great truth. I think, for the most part, people respond to something that they recognize in themselves or recognize in people around them, and laughter is a way of acknowledging that.

What is comedy structure, and how does a morning show go about creating it?

I think every morning show has or should have its own point of view, and I think it starts with characters. First, there's the main character, the host or the hosts, and then their interaction with the other people on the show.

So rather than think about jokes, think about who are the characters here and how can we interact?

I think the weakest shows are the ones that try to follow a formula, and I think the strongest ones are the ones that continually create their own formulas and create their own ways of approaching material and approaching ways of entertaining their audiences.

Some morning shows are fun, while others are funny. Is it better to be fun or funny?

Well, I think it's okay to be fun. It's better to be fun than be funny if you're not funny.

One of the guys I love the most in radio is Adam Carolla. I haven't heard his new morning show, but I used to listen to him on *Love Line*. Adam would go on these rants – he had his own unique viewpoint, and his opinions were couched in blunt observations. But his comedy didn't come from making fun of people. He was making fun of Adam Carolla being this nut.

Now Adam Carolla is a funny guy, but if you're not Adam Carolla – if you don't have a humorous perspective on something – make it fun. Because trying to be funny and not being funny is the worst.

I think it's better to enjoy yourself and have a good time and surround yourself with people who have interesting points of view rather than trying to do comedy bits that are kind of weak, because there's nothing worse than a bad joke, and a bad joke isn't balanced by one good joke because one bad joke is sometimes deadly. And don't push. The worst thing you can do in comedy is to push for laughs. Just back off, tell the truth, find what's odd about yourself, what's odd about how you see the world, and just share it. Don't think everything you say has to be the funniest thing in the world because it won't be.

Radio's Disruption Is Ready to Happen

An interview with futurist Watts Wacker.

Watts Wacker is a well-known futurist, CEO of FirstMatter, LLC, and author of many business bestsellers, including The 500 Year Delta, *and more recently,* The Deviant's Advantage. *Watts is described by* Wired *as "the inheritor of the Marshall McLuhan tradition," so media comes second-nature to Watts.*

Watts, what trends will matter most in the next few years that would relate to people who work day in, day out in radio?

One of the most significant trends is what I call "self-selecting social organization." People are looking to find people like themselves and coming together in almost a neo-tribal orientation of living. And there's a tremendous opportunity for all media, particularly broadcast media, to facilitate these people finding "themselves" in the easiest possible way. And it would also result in a lot of new business models for radio.

Like what new business models?

Well, I like to use the example of video podcasting. I know a couple of women in Nashville who are 23 years old. They video podcast a show weekly. They're suddenly getting people to give them 25¢ an episode. They do it every week. And if you get 50,000 people to send you 25¢ a week for 52 weeks, that adds up in a hurry. And suddenly, these women are their own production studio, and their job is just being themselves, where they podcast what it's like to be 23 and be a mom in the world today. They're putting together a neo-tribe of young women who are moms, and they're facilitating them obtaining information.

That's what I mean by a new kind of business model. That is why you see *Time* magazine saying the person of the year is "you."

Here's another example: It's what I like to call the art of the short view. This would be organizing and selling on a one-time basis something to hundreds of thousands or millions of people where the opportunity has a half-life of about three weeks, and then it goes away. But then another one comes and rolls right across the beach like a wave, one after another after another after another. And you learn how to put together groups of people and get each of them to send you five bucks in the period of a few weeks, and then it goes away, and you figure out the next one. And it's not that you do it once. You figure out how to do it, you know, 20, 30 times in a year.

Another important trend relates to what's happening with miniaturization in electronic components. If you look at Moore's Law, there are likely to be five more doublings of capability and halvings of cost. That means in 10 years you'll have an iPod the size of a pencil eraser and costing about $7-$8. Why wouldn't someone at that time consider just embedding it inside themselves? We embed chips in our pets already.

Perhaps the biggest trend that I would pay attention to in the short run is that while consuming is never going to go away, consuming as the defining criteria for individuals is. We are now using our media consumption as opposed to our physical consumption to explain who we are.

So you don't go to a party anymore and say, you know, "Where'd you go to college? What kind of car do you drive?

Where do you live?" Now you say, "What do you blog? What websites do you surf? Have you read the article in *Vanity Fair* on terrorism in South America? Are you an Imus or a Stern person? Have you seen *The Departed*?

Whatever it is, we are revealing ourselves through our media. We are becoming focused in life around ourselves as media. So today, "I am the medium."

You can see this play out in the creation of synthetic economies, where people are literally making a quarter of a million dollars in Second Life, which is a very popular emerging metaverse. They're making a quarter of a million dollars in the virtual world and downloading it onto their ATM card in this world.

And when that happens, suddenly you want to be in the broadcast industry and the content business, not just in the physical world, but also in the virtual world.

Where do you see the future of radio?

One of the reasons that I think radio could have such a vibrant next 50 years or so is the issue of imagination and particularly intuition and the allegorical composition of what radio represents. With radio, you fill in a lot of the blanks yourself. And I think that's one of the great hidden opportunities of the medium you really notice when you hear great radio, whether it's content or advertising messaging.

You use the expression "great radio," and I'm wondering what you feel great radio is, and what great radio will be in the next five to 10 years?

Well, my favorite closing statement from any broadcaster is "I'll see you tomorrow on the radio." You can fill in your own thoughts and develop the story in your own metaphors. That allows you to dig deeper into your own experiences, and it's one of the greatest hidden assets of the medium.

People are so conflicted by a world awash in uncertainty and complexity that allowing people to answer as if you're providing a question as opposed to giving a prescription is literally what the medium can do. Because radio doesn't have pictures, it actually becomes more of a benefit over the next 25 years.

Take an all-music station where someone was smart enough to realize they could buy it for four months and remove all of the advertising on it, which Snapple did in Boston, and you're touching that dimension of what I'm talking about. Even though it's a format of all-music, you're doing it in the way you're presenting your commercial messages as opposed to strictly "running spots."

There is a real opportunity for talk radio storytelling. You know, the only constant today in the world we live in is storytelling. And when you start putting forth questions instead of answers and you do it in a storytelling format, you could take talk radio to a whole new 2.0 – involvement with people.

I heard this on a student radio station for a local high school, which I listen to with great regularity, because these kids are figuring this out. If I was in the broadcast industry I'd pay a lot of attention to student broadcasters and what they're doing, because the alternative formats are being delivered and developed for you right in front of you.

And what were those high school kids doing that you can't hear on your local radio station there in Westport?

They were dealing with the tribe. They were clearly in a very micro-market. These kids knew that they had a different kind of audience, and they were talking to a tribe. They were not doing a format. They were touching all aspects of that tribe's interests as opposed to developing a format that could go to people who like that format on occasion.

And so they are holding their audience throughout the entirety of their listening availability. They're not channel-surfing radio anymore. They're staying with one station that has all formats rolled into one – for them and just for them.

You're describing a scenario where it's not one radio station with a massive audience. It's lots of radio stations digitally, Internet, whatever, with tribal audiences.

Correct. See, you're a genius. You said it better than I could.

This is why change usually comes from outside an industry: Because the leaders in the industry are afraid that change results in them losing power, market share, revenue, profits, whatever. All of the above.

And so the disruption is ready to happen, and it's really whether or not the radio industry is willing to realize that the pain of staying where they are will be greater than the pain of changing.

"Radiogenic" – Quality before Numbers

An interview with Norman Corwin.

There are legends in radio. Then there is the legend's legend. And that could only be the great Norman Corwin. Corwin was introduced into the Radio Hall of Fame back in 1993. This man has won Peabody Medals, an Emmy, a Golden Globe, a DuPont-Columbia Award, and he was nominated for an Academy Award. He's an influence on such names as Walter Cronkite, Orson Welles, Rod Serling, Ray Bradbury, and so very many more. He has worked in our industry, on and off, for most of a century.

Corwin may be an odd choice for a book which is mostly about radio's future, not its illustrious past. But everything radio is today has roots in that past. And Corwin is one of the most important figures ever to write for a microphone and easily the most notable person in radio I have ever spoken to.

Who better to link the past to the future?

Norman, what is your take on radio today, especially the world of talk radio?

There are these self-appointed oracles who distribute their prejudices, but that's not a very hallowed department of radio broadcasting. What I am talking about is the artistic end of radio. Its capacity to create programs that were, let's say, "radiogenic." Programs that were suited for the medium and which expressed the individuality and characteristics of broadcasting to what is essentially a blind audience, an audience equipped to listen to the content, not to be distracted by hairdos or low necklines.

I'm afraid radio will forever be now on the outskirts of broadcasting rather than the center of it. While it was going, while it was flourishing, it did produce some work of consequence, and I'm very happy to be included in radio's so-called "golden age." So, what we're talking about is a kind of nostalgia, but my regret is that many others who where as well equipped or better equipped than I to exercise the art of radio have been prevented from doing so by the exile of a once very lively and productive medium. The modest budgets (by comparison to television) are no longer available.

What was it like to create programming in that era and know the impact that this programming had on people?

Well, I would sometimes get letters that were better written than the programs they were writing me about. I had a very good audience. And the CBS Network gave me freedom, and without that freedom I don't think I would have been of sufficient interest to occasion this interview.

You talk about the freedom to create the content you want. With all the talk today about the primacy of "content" and its role in the new media universe, how important is freedom to create in that context?

Well freedom, of course, depends on what is done with that freedom, you know. Freedom has limitations. For example, freedom of speech does not permit the crying of "fire" in a crowded theater when there is no fire. A dictator, for example, is free to be cruel, and that's not the kind of freedom I'm talking about. I'm talking about the kind of freedom that ennobled the birth of America. No country was ever established, no government was ever established, on as high a level as our own with the Declaration of Independence, and that's the kind of freedom I'm talking about.

Let me give you a single example to fill in all the spaces. When World War II in Europe was about to end, and Nazi

Germany was on the ropes, the program manager at CBS came to me and said, "Would you drop what you are doing and prepare a program to be on the air on the night of victory in Europe?" And I said, "Certainly, I will do that." And I did. I dropped what I was doing and prepared. Now, the CBS Network was going to requisition a commercial hour, take it away from the sponsor in prime time and give it to me for an unsponsored hour commemorating the end of World War II in Europe. And they never said to me, "Let us know what your approach will be." "How much will it cost?" "Who are you going to cast?" And "Can we see the first 20 pages?" They said nothing of that kind; they left it to me.

And with that freedom, I went on to do a program that is, to my great surprise and pleasure, repeated annually on the anniversary of VE Day – it reached its 60th anniversary this last May. So, that program could not have been written and broadcast without the freedom that was given me, a freedom no longer obtainable, I would have to say with sadness, at CBS or anywhere else. The idea of a program, an hour, prime time, going on the air unsupervised without being checked by the office of standards and morals is unheard of today.

It was unheard of five years after it was done.

Norman, I think a lot of the people working in radio today would love to have the kind of impact on the lives of their listeners that you've had on the lives of so many.

Well I was very lucky to come along when I did, and lucky to have a program, authorities, executives as freedom-loving as I was.

What advice would you give to broadcasters today who want to make the kind of difference in people's lives that you were able to make?

To go after quality before they go after numbers, and I believe quality will attract numbers given an even break in the market.

Do you see that quality out there today?

No, I do not, and I'm worried about what will happen with the newest phase of radio. And I'm a little worried that when $500 million are put on the line for Howard Stern, the prospects are not all that bright.

Maybe the question is, "What do they do with the rest of their money?"

I have to say that radio is far from defunct because it continues to be a profit center. But what was the outstanding broadcast of last year or the year before? There were such highlights when radio was in its full stride.

What's the most important lesson that you would convey to broadcasters looking for advice from someone of your stature?

To return to the concept of freedom of expression, to hold sacred as all of us should do the First Amendment to our Constitution and the freedom of speech, and to encourage at an early age the listening to radio.

You know, there was no phrase in all of radio's history equivalent to the term "boob tube" or "couch potato" because radio is here for the ear and the mind. Radio has always encouraged collaboration to the way reading a book is a collaboration between the reader and the writer.

Bearing that in mind, the way to assure quality in broadcasting is to promote freedom of expression and to back that up with a few dollars to pay the artists.

Do you think the industry today has lost sight of the opportunity created by that collaborating between radio and the ear?

I think it has very badly lost it.

How do we get it back?

Application and the determination of a few people and a friendly government, a friendly FCC.

Where are the poets today? In my time, Archibald McLeish, a Pulitzer Prize winning poet, wrote for the medium, and so did Steven Vincent Benét, and so did Dylan Thomas, who first reached world recognition from a radio play that he wrote called *Under Milk Wood*.

These are the storytellers.

Yes, and it all begins in a cocoon of freedom.

"Radio, Get Your Head Out of the Sand"

An interview with Joseph Jaffe, advertising and marketing guru.

Joseph Jaffe is the author of Join the Conversation: How to Engage Marketing-Weary Consumers with the Power of Community, Dialogue, and Partnership. *He is the president and founder of Crayon, LLC, a new marketing innovation company. He is previously the president and founder of Jaffe, LLC, and a director of interactive media at TBWA\Chiat\Day and OMD USA. Clients have included Coca-Cola, American Airlines, Starwood Hotels, Proctor & Gamble, K-mart, and many more. He's also author of* Life After the 30-Second Spot. *Joe also does a very popular podcast at JaffeJuice.com.*

In your book, Joe, you say "Conversation trumps communication." Tell me what that means.

Even when the entire media industry is at the top of its game it has become impossible now to break through the clutter. Without conversation between your brand and your audience, without viral or word-of-mouth efforts, you can't stand out from the crowd, break through that noise, and ultimately deliver a return on investment to shareholders and specifically build your brand.

Radio, TV, and print are considered "traditional media vehicles." How are these different from the digital and interactive tools of the conversationalist?

Well, you ask the question in a very interesting way because I would also say that a lot of digital and interactive online advertising is as much communication-focused – not conversation-focused – as anything else.

I remember a strategic planner that I used to work very closely with when I was working on Madison Avenue, once asked the question, "What is more interactive: A coloring sheet on a McDonald's tray table or a banner advertisement?"

I think it's probably too easy to say, "Television, radio, and print are the dinosaurs and the outmoded forms of building brands or engaging consumers." It's too easy to do that. That said, these more established media have to work so much harder to stand out from the crowd and go that extra mile.

Facebook is a great conversation tool, but the ridiculously oversimplified skyscraper ads there don't impress me at all. What impresses me is any form of communication that gets you talking – that can at least transform into conversation.

The tools of conversation ultimately are community, dialogue, and partnership-based. And so, a spark becomes a raging fire of passion, in a sense. It is the ability to extend and enhance and amplify a message into an event or an experience or something truly transformational.

Suppose I am a radio station and I want to get deeper into the conversation business. Focusing on the listeners, what are some ways that stations could dive headfirst into conversation?

Radio is the one form of big media which has probably been least innovative when it comes to adopting and embracing some of the new approaches that arguably have threatened it the most.

I'm not necessarily saying that the television industry has done a very good job at all, but at least they're experimenting to no end. A lot of shows are trying to keep viewers engaged by using SMS technology. On *The Amazing Race*, the contestants that finish last and leave the show are actually sequestered into a house and stay there until the end of the series, and there is exclusive content available online to connect with these past contestants.

In the print industry, online versions of newspapers have embraced social media quite extensively. They've embraced pretty aggressively certain conversational approaches, such as the ability to comment on an article, to digg the article, to forward it to a friend, etc.

Now I'm not really sure what the radio industry has done at all to shift their thinking, their mindset, their philosophical approach to their art and their trade and their profession, to fully adopt and embrace the elements of community dialogue and partnership.

One might argue that radio has always been in the business of dialogue in terms of welcoming callers, but I think there's a difference between taking live calls and being listener-driven like, for example, podcasting has proven to deliver.

We've seen, for example, how newspapers and magazines have brought in a lot of comments or commentary to their content and into their text. I think there's a huge opportunity for radio to do the same, for radio to figure out ways to bring the audience into their programming lineup or into the content itself to fuse these two different worlds.

So maybe I should throw it back to you and ask you, have you seen much innovation? Would you agree with my statement that the radio business has not been overly innovative at adapting and in so doing adopting some of these seismic shifts in the landscape towards consumer-generated content or citizen journalism?

Yes, I agree. I think that radio views itself in the business of selling spots and everything is evaluated against that metric. There's little understanding about how getting deeper into podcasting, community, and conversations might relate to our ability to maintain audiences and sell spots.

Okay, you raise a very interesting point, because you've zeroed in on the business model, but a business model that is tired and economically does not make sense anymore.

The clutter right now – eight minutes sometimes – of advertising that is not relevant, advertising that is – from a creative standpoint – beneath sub-par and inferior. It's bizarre, especially considering the radio listener is probably the most empowered consumer because of the ability to easily change the channel. Relief is literally just a finger's distance away.

So here you have an industry that doggedly continues to "spam" its customer, its listeners, instead of figuring out new revenue streams, and/or business models and/or smarter, more considerate ways of engaging its listener base and its community. And, by the way, there is no community at the end of the day; it's just a bunch of disparate listeners.

On one hand you've got this incredible potential, this "theater of the mind," but on the other hand you have content that has not yet been liberated and become more mobile, as it should.

The business model is in dire need of an extreme makeover.

What kind of content is on radio in a form that suits some of these conversational tools?

I see this incredible wealth of untapped talent out there in every station's audience.

Today, with a program like Garage Band, a Neanderthal can become a musician.

And then you've got this rise of citizen journalism and podcasting. In one of your older blog posts you mentioned only 30% of the U.S. market has even heard of podcasting, but you know what? It's a storm that is coming. Soon every single new car will have a neat little spot to plug in your iPod or your MP3 player, and those MP3 players will be WiFi-enabled. And suddenly people will be calling in live to podcasts as they're being produced through the power of WiFi. And when that happens, the radio industry could lose.

The radio industry could lose on so many different levels: On the level of content, on the level of commercials and clutter, on the level of control, on the level of community; I can keep going on and on and on.

So my message to the radio industry is, "Get your head out of the sand and innovate or you will die."

And that's not me pointing a finger saying, "You're dinosaurs." It's me saying, "You guys need to innovate and take a good long, hard look at yourselves and figure out how to evolve. Because this is about evolution and about revolution, and right now I think the radio industry is losing on both counts.

Well let's make Joseph Jaffe king of the radio world, and you have your own station to play with. What are your first steps down this path?

The first thing I would do is to begin fusing the two worlds together. Why, for example, is there no Yankees fan radio station? Why is there no official Yankees podcast out there? Why are we not localizing and regionalizing and doing a much better job at being able to bring radio content to consumers on a more targeted, relevant, micro-geographic level?

Radio has to stop thinking about itself as this communication-based channel that lives inside cars and stereos. I remember listening to Janet Robinson from *The New York Times*, maybe two years ago. She said, "We are not in the newspaper business anymore. We are in the content business." And I think *The New York Times* has probably done a fairly good job at proving the fact that this wasn't just talk. The fact is, *The New York Times* is as much in the television business and the radio business as it is in the newspaper business. In fact, *The New York Times* is in the news business and the content business as opposed to restricting itself to any one particular channel.

In your book you write, "There's no question, the very definition of 'media' is changing quickly. In fact, it may behoove us to abolish the term completely in favor of the more familiar term 'content.'"

Exactly.

Because content is king...but what if there's a revolution that does away with the monarchy? I've played around with this idea that content may be controlled by the monarchy or the masses, the peasants, if you will. That's one way to think about consumer-generated content.

When we think about radio, we need to say: Wait a second, are we in fact talking about audio content? Not just radio content, but audio content? And why are we restricting ourselves to audio? Because around that audio there surely needs to be every other form of multimedia from video to text to photographs, and so on.

So the radio industry needs to figure out what their content is and how to monetize it. Because there's just no way that consumers are ever going to sit through eight minutes of irrelevance again. There's just no way. And I think the radio industry is fooling itself if they believe people are not only sitting through this, but literally converting and responding and purchasing product.

You know, if that old John Wanamaker statement is true, "50% of my advertising is wasted, I just don't know which half," then I think in the radio industry it's probably closer to 80%.

Radio is increasingly being pitched as a "reach" medium. But in your book you argue that's old-school thinking, that accountability – actually delivering response – is where the ad industry is heading.

Right, I use three words: Reach, connect, and effect.

Radio and print and television have their business models built around a reach-based methodology. But just because you can reach people doesn't mean you will, and if you do reach them, will you connect with them? Will you engage them? Will you be memorable for all the right reasons? And even if you do that, will you be able to effect some kind of an action, whether it's bringing them closer to the brand, signing up for an email, requesting a brochure or a direct mail piece, or some kind of conversion or transaction?

You also write that "people are the message."

Absolutely, people are the message.

People deliver the message, and they are the message as well.

I use the example in the book where I listen to *Daily Source Code* with Adam Curry, and he kept talking about his great experience flying Virgin Atlantic. He was talking about Virgin because, you know, he's a human being and he loved the experience. Well, guess what? When I was flying to the U.K. and thinking about which airline I wanted to fly, I ended up flying Virgin – all because of one person – one person's not even overt recommendation, but endorsement. I've never met this person before, but I trusted him. I just felt like I trusted him. I trusted him because I listen to him regularly. Gee, that sounds pretty familiar.

It sounds like the radio business.

We trust these people we listen to because of the power of asynchronous intimacy. We build a relationship and rapport with them over time. And that trust almost goes out the window in the radio business when I hear, for example, these Dan Patrick live reads, you know: "Hey, I'm Dan Patrick and when there's a snowfall in the…(laughter), I use some brand of gloves, or whatever." And I'm like, "Come on!" It's ridiculous.

People are the message. They're carriers of the message, but they're also originators of the message as well.

So, you know, there's no ad campaign any more. There's no start. There's no end. It's a very fluid picture. There are no more sellers. There's just life, you know? Life happens. Life is around us. Humanity is around us. Passion and engagement are natural. It doesn't belong to the media companies, and it doesn't belong to the brand marketers. It belongs to us, and I think that's the opportunity.

Radio has to figure out how to bring all these different worlds – "church and state," content, commercials, trust, influence, community – together to create this incredible mash-up.

And when they do, they may find themselves enjoying a new lease on life.

Radio Trend Spotting 2011

An interview with author and marketing guru Richard Laermer.

Richard Laermer is president of RLM Public Relations, a columnist for the Huffington Post, *and the author or co-author of numerous marketing bestsellers, including* Full Frontal PR, Punk Marketing, *and his latest bestseller,* 2011: Trend Spotting for the Next Decade.

Richard, why did you write this book?

Well, right now, we're in this decade of mediocrity. There's nothing going on – everybody knows this. It's just a ton of stuff happening, but no bolts from blue, nothing to connect. I wanted to look ahead, just like I did with *Trend Spotting*, the first book I did on this subject in 2002. But now I wanted to show people how they could take a deep breath and start the new decade, which I call 2011, with a clear head and a positive outlook and realize that we can learn a lot from what we went through in this unnamed decade.

So give me a sense of what some of the most critical trends are. What do you think are the core themes of the next decade?

Well, for one thing there has been a kind of inflexibility in the air – people have their arms folded, and they're always looking for somebody else to clean up their messes; it's always somebody else's fault. People lack the power to get things done.

I invented a word in this book called "Gumby-tude," which is based on the green character, Gumby – not the one Eddie Murphy played, but the real Gumby.

And the idea is, "Gumby-tude" is utter flexibility. People will realize that they're responsible for "getting it done," by hook or crook, as the old saying goes. And I think people are going to learn to look to themselves. And what I'm saying is, look, let's not pay attention to all the great futurists out there. Let's pay attention to ourselves. Let's go out and find the trends that we need to know about so that we could either make money or make more money. I want folks to learn how they can be their own trend spotters.

My new book is a guide to "do-it-yourself trend spotting," if you will.

In the radio business today a lot of people are waiting for the leaders at the top to do something that makes a difference in their everyday lives that is a positive difference rather than a negative one. I don't get much of a sense that people are feeling this self-reliance that you say will be critical for the next 10 years.

Radio, which has been around forever and will probably be around forever, has a problem: There aren't a lot of people who are willing to take risks.

In a lot of the work that I do, I'm always trying to get people to think about what they're not doing, and whether or not they're just doing what they're doing because they were told to do it.

Radio has to think about the way it communicate its message. Radio is stuck in this thinking of, "Well, listeners will have to pay attention because we're repeating our messages over and over again." You and I are both marketers, and as we both know, communication is everything. And there's no worse communication than a kind of stodgy group of people who are just going, "Huh, what do we do now? I don't know. Let's talk about the greatest hits of the '50s, '60s, '70s, and '80s."

You work across many industries. Is radio different from some of these other industries? Or is radio typical of an industry this big and this mature?

Well, it is different. I work with a lot of technologies. What I find is that radio seems to always be fighting a battle against somebody, as opposed to saying, "To hell with you guys. We're gonna do things our way, and we're gonna change."

In the last 15 years or so, radio has felt beaten down by so many other types of radio-like experiences, like satellite and Internet radio, etc. And I think that radio always feels like they have to be on the defensive. And that's tough – that's a tough way to market, when all you do is think about how, "Oh, God, we're so old fashioned. We've gotta do something." It's not a real strong way to fight for customers.

And I think that most people in radio are not thinking about how to create the radio fan – not the new listener, not the person who doesn't listen to radio – but how to get their current customer to think about radio as being that one thing for them – that one communication device for them. As opposed to all this craziness where people are marketing this HD stuff and trying to get kids or people who have just dropped away back into the fold.

When I turn on commercial radio, I feel like I'm listening to the same promotions and ideas that I've been listening to my whole life. You know, I mean, we're evolving human beings.

So give me just a couple of the ways in which you advise people in the book to spot trends, as opposed to fads.

Well, for one thing people have to be much more alert than they are today. Thanks to the Internet, most people use a kind of personalized way to get their information. They only go to what's interesting to them, and that's really a failing.

The real trend-spotter doesn't do that. Being interested in something, only those things, doesn't make you an interesting person. And in order to be a magnet for information, you have to know a little bit about a lot of things. But the truth is, people today only know about the entertainment world or they only know about radio, or they only know about sports, or they only know about yoga. But they don't know about a lot of other things out there. And you have to be that person that people want to connect to because you know what's going on.

So to be a trend-spotter, get a little bit of everything from everybody, and pull it in.

I'm always surprised, Richard, that in radio, for example, folks will go out of their way to go to a radio convention, but will rarely attend any gathering focused on new media, when that is the industry radio is now a part of.

Well, people don't think about what their customers do. They think about their own industry, but they don't think like their customers.

Radio stations are doing less research today, not more. Why is that? That drives me crazy. Why aren't they looking at what their customers want out of their own lives, you know?

Just because your station has listeners doesn't mean you're connecting to them. In the book I talk about auditing and how you can find your listeners, users, or whatever, and get to know them. I mean, it's so easy to do that now. And I mean real audits – ask the hard questions, like "Why do you hate us?"

Let's talk about the recent re-branding efforts of radio, which I know you're aware of. When you've got something like radio, an industry that big with that many tentacles, what is the best way to "re-brand" it, if that's even the right thing to do?

Well, I saw the re-branding of the logo and the slogans. It's yet again an example of people not realizing that this kind of wasted money never really helped any industry. I mean, it's been a long time since "got milk," you know.

If you're going to re-brand radio, you have to do it in a way that makes people feel that it's not a slogan, but it's real, that something is happening – there's a revolution brewing that the listener wants to get involved in. And you have to do that by getting the listener involved with the re-branding effort.

It's definitely not about slogans and logos, that's for sure, because I saw the slogan and the logo, and it's not good. It has to be a lot messier than that.

I've said this to you in the past, but I'm always surprised about how little radio stations get really involved in their communities. And I don't mean sponsoring softball teams and going to the local bakery, but I mean really getting involved – taking over whole neighborhoods and commandeering streets and just showing people that – you know, these are the public airways, right? It's not just "Yes, we have the airways, and yes, we keep changing our call letters. And we hope that one day you like us." Instead, "We commit to this, and we commit to you. And we're going to stay on this line of thought because, like any good marketer or seller, we believe in the mission of who we are."

Mark, we have known each other and been friends for a long time – and I never see radio commit to anything. Most consumers and listeners are so smart – remember, these are the people who invented hype, so they know when they're being hyped to death. And radio is a victim of its own hyping – not its own hype, but its own hyping – it's constantly saying things, constantly changing, and it's like a whirlwind. People roll their eyes and are like, "Oh, look, that radio station is now X instead of Y. Big whoop!"

What you seem to be describing is a "re-branding" that comes from the inside out rather than from the top down or the outside in by slapping a new label on something.

If I ran the world, I would put all the branding companies out of business because I don't really think they're actually helping people. I mean they make a lot of money, and they say a lot of cool things, and they do great PowerPoints, but it's a joke now.

Branding is about connecting with people and knowing what you want to say to them. Pandering is deadly, and I see radio doing a lot of pandering. But if you're actually resolute about who you are, then prove it without a shadow of a doubt. Do what you think is right, because you're the station that wants to be different and wants to prove its value, and you promise you're going to stay like that.

Put it in blood!

The New Business of Radio

An interview with Triton Media COO Mike Agovino.

Mike Agovino is president/COO of Triton Media, a sales and marketing company specializing in providing digital tools, platforms, and sales solutions to today's radio industry. Mike is a former president of Katz Radio, he ran Clear Channel's national sales company, and he was co-president and co-COO of Interep.

I think it's safe to say based on your background that you have a pretty good awareness of the business of radio from the top on down. How do you assess the state of that business now?

I think the industry is faced with an incredible amount of challenge on all sides of the business. I mean, you wake up this morning, and you see that we're in Washington battling royalty issues on both over-the-air and streaming. We're facing fairness doctrine issues. We're looking at Philadelphia and Houston ratings that literally change the landscape , and I think we're looking at a time where reinvention across most of the traditional job descriptions in radio is necessary.

I think the role of program director really becomes the role of brand manager, and that person that used to worry about getting a number in an Arbitron book once every three months through that methodology is now focused on managing a brand and all the different ways that that brand morphs through all the different distribution channels. On the sales side, the same type of reinvention has to be done.

For the most part as I look across the landscape, we're doing the same things we were doing 20 years ago. We're selling spots the same way we were selling spots 20 years ago. We're using the same technology in all but two markets, the same measurement service, and I don't think the skill sets of our salespeople have evolved a whole lot, and that's the thing that I think is most critical in the sales side of the world.

We are not a technology company per se. We invest in technology companies and help to connect those back to the sales process, but we're primarily a sales and marketing company, and we look at what's happening in radio today inside these sales forces and see sales staffs across the country that are not prepared to bring to advertisers the type of marketing opportunities that radio needs to bring to advertisers today.

They don't understand the digital world. They don't understand how to integrate it into proposals with on-air. The stations have not invested in their digital future and brought the type of extensions and new distribution platforms that are necessary. They've not invested the money in the website that's necessary.

And you sit and you look at Philadelphia or Houston numbers and recognize that the primary currency that we've utilized to produce revenue is going to tell the advertising world we've got 20, 25, 30% less tomorrow. We had better be coming up with ways to increase the value of that by 25, 30-plus %, or we're going to be in trouble.

You're talking about the impact of PPM specifically in AQH rating, right?

Yes. You read a lot about what Arbitron's trying to do in terms of educating the agency community and the stations, but you look at a Houston ranker today and see a radio station like KODA, which has been the market leader for a very long time, and you wake up one morning, and your 25-54, 6A to 7P rating is cut in half.

It's great that your cume doubles, and that you're reaching a much more significant portion of the marketplace, but we're the ones who created this statistic. We're the ones who really built up in the agency world the idea that the quarter-hour rating or the "impression" was how they should build campaigns for the medium.

By the time all the new PPM measures translate to that buyer, she buys average-quarter-hour ratings by looking at the top of the ranker and working her way down. If I'm on that sales staff of KODA, I'm used to being on the top of the ranker with a nice, fat rating, 6A to 7P, and I wake up today, and I have half of that rating, and I'm no longer on the top of the ranker.

That's dramatic change, and managing through that change, and providing an advertiser or an agency with a rationale that allows them to pay the same amount tomorrow or more than they paid yesterday while the statistics say I have half as many listeners on average, that's a big challenge.

You were saying before we started this conversation that looking down the list of AQH ratings now, compared to what it was in a diary world, almost simulates what we would have expected to happen 10 years down the road with the march of technology.

Well, it is. We talk about all the challenges of satellite radio and the iPod today, and the talk of Wi-Fi and WiMAX tomorrow, and a world with thousands of options in the car, and we always talk about radio being challenged with a flat revenue scenario today, and sometimes we paint the picture that with all these additional options and all this technology out there, we could be flat or down over the next 10 years.

You look at one piece of paper that says the market-leading radio station has half the audience it had yesterday, you know, that impact could come a heck of a lot quicker, and to me, PPM just completely reinforces the idea that we've got to create a national digital model that activates across the traditional broadcasting industry and primarily radio, which is our expertise.

Because if we are not making more money off the air than we are on the air 10 years from now, we're in trouble.

And I think today most of the broadcasters are very pleased if they can say that they're producing 2% of their revenue through alternative platforms to the traditional on-air commercial, and we know there are a lot of people who are doing a lot less than that, but you look at these numbers, and you look at everything that's going on technologically, and I think broadcasters are thinking too small – they ought to be realistically expecting that they're going to do more revenue off the air than they do on the air a decade from now.

And you really think that's possible? That more revenue will be generated 10 years from now from online, from other so-called "non-traditional revenue" – or as I call it, "new-traditional revenue"?

Yes. If I put it into a bullet, I would say that we believe that the opportunities presented by new technologies outweigh the threat of those new technologies, and I look at radio and think, "God, we're the original interactive medium."

For years and years and years, we've taken song requests and done contests and had events and interacted with our audience all day long, and 99% of the time the listener calling in for the contest got a busy signal, or couldn't get the tickets to the event, or wasn't able to show up at the appearance or the van hit to meet the street team, etc.

And now with all these new technologies, we have the ability to let everyone in the audience participate in every single thing we do. There's a huge opportunity there. I look at everything that's going on in the world of social networking and think, "Gosh, that should really exist under a radio station." And what are the viral syndication possibilities of social networking that exist under a radio station's brand that allows it to go sub-local?

I was talking with a friend who runs some radio stations in Phoenix the other day that are news, talk, and sports, and the ability to take a listener who might also be the coach of the Glendale Little League in Arizona, and have them uploading video of the finals of the Little League Playoffs, and then allowing 50 friends who may not be listeners to participate by viewing that and uploading their own.

And you can see the day not too far off where you actually grow your off-air audience to larger proportions than your on-air audience if you're using all of these tools the right way. And so, I believe in the additional dimension of video, I believe in the potential of personalization and everything that can go on underneath the brand that I may have grown up with or I love today.

I mean, if I am a hardcore loyalist to a radio station, the station now has a lot of opportunities to take advantage of that outside of just the listening in the office and the listening in the car. And so, yes, I think that when you look at all of that and look at all of the different ways that you can interact with that audience and monetize that interaction and present a better value proposition to the advertiser than you do today, then absolutely we could be making more money.

But if we're not thinking that way, then we're probably not committing to this strategically, and I think too many companies have not made a real commitment – the kind that goes beyond what you have to say to keep the analyst community at bay or to keep your staffers thinking you're paying attention to this.

I mean a real commitment: A program, budgets that exist down to the individual account executive level where there's an expectation that everyone is going to produce "X" amount of revenue, that they're given the right tools, the right training, that the company has invested in the right technologies to allow the radio station to do all those. I don't think it's an option. I think it's an absolute necessity, and I think across the industry right now, we've got a lot of people going through the motions and not really committed to it.

There's also a lot of confusion out there. There are a lot of alternatives. There are a lot of choices. There are a lot of paths, and nobody knows which paths lead to a dead end and which lead to a pot of gold.

One thing we learned was that the decision-making process combines program directors, promotion and marketing directors, general managers, group programmers, VPs of digital, COOs, and CEOs.

There are so many new-to-the-radio-world vendors out there. Twenty companies in streaming. Twenty companies in mobile. Twenty companies in database, and to literally maneuver your way through all of these vendors and actually have criteria where you could separate them in terms of technology, functionality, etc., is a task that no one in the radio industry may be qualified to do. I know I'm not.

It's just too much to sort through . So one of the first big determinations we made is that all this has to get made a lot simpler for the broadcaster, and that's why at Triton, in partnership with our good friends at Mass 2 One, we've launched [eco]; [eco] is a complete digital ecosystem for radio.

What we did with all the Triton partners was to offer a completely integrated suite of applications tied to that database program where we could then go to radio and say, "Look, you don't have to entertain 20 streaming vendors, and 20 mobile vendors, and 20 social networking vendors, and 20 database vendors."

I'd like to think in the case of database and streaming, we have partnered with the best companies in each of those individual categories and integrated them so you've got a great solution working together.

And so, I think we've made it significantly easier on both sides to allow the broadcaster to make a decision more easily and to allow the listener to have a much deeper, better, stronger, easier experience, because they log-in once and are able to access everything.

Ultimately, something has to be sold and something has to be bought, and what's the strategy for maximizing the revenue from all that?

Well, let's take a real life scenario up in Canada. Mass 2 One's got a partnership with Rogers Media up there. I think it's 54 radio stations today. They launched the first one about a year ago, and they are up to about 57 million page views a month for that group of radio stations, and that's a good mix of markets. So you figure they're averaging about a million page views a month in 2007 off a broad mix.

I can easily see a day in the not too distant future where we've linked several thousand, not 50, radio stations together averaging many times what's happening today when the technology is all introduced the right way to the audience and the engagement factors are what we want them to be.

You're talking about billions of impressions a month across an aggregated network that is that big. You're talking about tens of millions of dollars a month in digital ad sales revenue by presenting a true vertical radio experience connected across those thousands of databases that can be diced and sliced a gazillion different ways.

The goal is to show up on P&G's doorstep and offer a thousand radio stations that have a very meaningful female 25-44 listener base. We place a survey within their databases asking whether they are planning to have a child in the next year, or are they expecting a child in the next three months, for example. If so they might receive some kind of special offer targeted right to them.

You can take an automotive company, a travel company, a financial company and integrate them into a database that's large and make special offers back to that group.

Look at what American Airlines is doing in their frequent flyer club. Look at all the different ways they're making money off of my loyalty to them. All of that can happen in radio if we activate it properly.

I came out of national sales in radio, and one of the reasons why I left it and we launched Triton in 2005 was because we saw a great void in where radio was going digitally.

We didn't see any of the traditional national spot companies making the necessary investments on the digital side to provide a national sales channel for radio's digital inventory, and when you look at the small fraction of radio's revenue that's coming from digital sources, almost none of it is national – it's virtually all local.

So our vision is to build a digital syndication company/ rep firm for the 21st century, and we're well on our way.

Given the difficulty stations have had to date developing digital tools, why wouldn't we expect them to have just as much difficulty selling the audience that comes to these tools?

This is not a fast build. That's why we spend a lot of time in each market meeting with the sales and programming staffs and doing extensive training. We invite the sales staff to bring in advertisers for two days, and we do sessions with a handful of local advertisers, showing them all of the different ways these applications can be used and how they compliment an on-air buy or simply provide a digital presence.

You know, even the average "C" student inside the average radio station who fell into this world of selling spots looks in the mirror at night and says, "Am I getting better? What am I doing to expand my horizons?" There's a lot of technology out there that will allow an agency to hook up with a pool of inventory, determine a price, and put a spot on the air.

I think the majority of time radio is bought, not sold. It's pre-planned to be bought. The criteria are very straight. It's simply negotiated. Right now we've got human intervention in that negotiation. I suppose because of the promotional nature of the medium we'll always have a certain amount of human, but we certainly have a lot more of it than is necessary.

And if you're that account executive in that marketplace looking at a computer that could basically do your job if the broadcaster was predisposed to let that happen, I sure as heck would want to be learning everything I can about all these new toys, and how I become an expert, how I convert my customers over, and how I build cutting-edge, integrated campaigns that utilize radio on-air and online.

So the radio business isn't *just* the radio business anymore?

The radio business is *not* just the radio business anymore.

For the companies that go down the same old path, not believing that the evolution and change is going to catch up with them, time will prove them very wrong.

I mean, my 17-year-old son spends half his life on Facebook, utilizing the 10 widgets that he's dragged and dropped into his Facebook page. Radio is not a significant part of his life.

And maybe it never will be.

But if it can, it's only possible if we're utilizing all this technology to make our offering to that demographic a lot stronger than it is now.

These digital solutions are becoming so much easier to use, and so many more choices are becoming available. It's no longer that futuristic to think of Wi-Fi and WiMAX, and thousands of choices in the car – most of them out of the radio industry's control.

So if we're not doing things to make sure we remain one of those primary choices, we're in trouble.

Your Radio Station Needs to "Zag"

An interview with Marty Neumeier.

Marty Neumeier is president of Neutron LLC, a San Francisco-based firm specializing in brand collaboration. He's also the author of several terrific marketing guidebooks, including The Brand Gap *and* Zag: The Number One Strategy of High Performance Brands.

What is a Zag and why do we need to do it, Marty?

All of us are facing a super-cluttered market and a world that's moving faster than ever. Innovation is leapfrogging all the time. So in order to keep up with that, in order to win in that environment, you have to not only differentiate yourself, you have to radically differentiate yourself. So I called that a Zag, a radical differentiation.

If you look like everybody else, if you're a me-too brand or a me-too product or service, you don't matter. This is really the point. How do you matter to your audience? What makes you different and compelling?

That sounds like the basics of positioning and branding. How are you taking that to a new, deeper, more edgy level?

Well, I think you're right, this is positioning. It's basic positioning. But I think what people need to understand nowadays is that when you're inside a business it seems like you're different because you're so close to it. But from the outside, it may not be different at all. I think people underestimate how different they have to be in order to make an impact.

And when they really confront what it takes to be very different, extremely different, they get nervous about that. What's going to happen to their business?! You know, why should I do something that nobody else is doing?! Isn't that risky?! Because that's the traditional way to look at business. If nobody's doing it, I'd be crazy to do it. And I say if nobody's doing it, it might be brilliant to do it. But how do you make your company safely innovative?

What you're looking for is "white space." Where can I go where no one else is so that I really stand out from the crowd? Even if the audience is narrow, it may be very deep for you. It may be an audience that you can create a lot of loyalty with.

If I'm a radio station and I'm committed to be dramatically different the way you say, how do I go about building a Zag? Where do I start?

Well, we like to start by looking at the brandscape. In other words, who do you and your customers consider to be your competition? Make a list of those. And then make a grid. On the left side you would write your competitors and across the top you would write the key features that drive your success or the success of this category. And you will see how strong each competitor is in each of those areas, and you'll see where the white space is.

So the big question is, where are people not competing? What are they not doing, not doing very well, or not doing very much of? Further, what are the players in this category not doing at all because they haven't even imagined they could do it? Or what are they doing so much now that if we did the opposite, we would create a whole new category of success?

You need to explore what's missing in the brandscape, what you have a passion for, and what you're willing to take a risk on. Exaggerate those and stop competing in the areas where the competition is too intense.

Okay, I see a couple of the difficulties. One is that the white space is where the risk is at its maximum. And the other is that what's popular and what's different can be very difficult to find. How do you deal with these two problems?

You want to be different and good, because that's where you're going to create new market space that didn't exist before and has a nice long run before anybody copies you.

There are lots of examples of this in different fields. In TV, *All in the Family* created a lot of white space. Just a whole different look and feel and attitude at the time. In cars, the Prius I just bought was good and different and people looked at it and said, "Well, that's a really weird looking car." But I guess, you know, it might be okay, since you get 50 miles to the gallon.

Or the Aeron chair, which everyone thought was really bizarre when it came out, yet at one point 30% of its manufacturer's revenue came from that single product.

So all these kinds of products started out baffling people and getting very strange feedback, but they also were perceived as having some value. And so that creates a kind of recognizable pattern that you look for when you're trying to find that white space.

When you expose people to a prototype of your product, what you're looking for is a response like: "That's weird or different. Hmm, I don't know if I agree with that. I don't know if I would buy that, but I certainly see the value in it."

When you get those two things together – different and good – that's when you start to think you may have a home run, and then you need to figure out how to de-risk it. And one way is to start small. Invest in little tests so you're not fooling yourself into thinking it's a winner when it's really a loser.

In radio we're dealing with multi-million dollar stations and obligations to owners and Wall Street and so on, and when you flip a format, you're making a wholesale switch where the risk is huge. But could we perhaps conduct these experiments online?

Yes, maybe try it online. Maybe start a brand, a separate brand, that won't affect your main brand if it fails. And if it looks like a success, then switch over. Often what happens is the big successes won't come from these large brands and large companies. They'll come from the small ones who have nothing to lose and will do anything to leap ahead of you. So the market is always moving faster than any one company. That's the danger.

What's really risky is to stay the same. So you need to keep up with the speed of the market, and to do that you must think out of the box, try little experiments on the side, perhaps. You don't have to spend a lot of money, but you need to prototype your ideas, build up a small audience to the point where you see it could grow. If it looks like you're going to ride a trend, so much the better.

And then you flip the switch.

For long-running and established radio brands, how do you strike that critical balance between novelty and consistency? Especially when a station is doing well today by not changing?

One thing I've observed is that long-running brands tend to do best right before they go under. Companies sometimes get so good at what they're good at, they forget why they got in business in the first place, which was to innovate.

And so they get this sort of cultural lock-in where they just can't change because those processes are just so ingrained, and there's nobody in the company who really wants to change, so they don't. And eventually they just go through that sunset effect – out they go and something fresh takes over.

You need to look for ways to stay young so you don't face that sunset prematurely.

Don't be rigid.

Kill Your Radio Station's "Sacred Cows"

An interview with Beau Fraser.

Beau Fraser is co-author of the new business bestseller, Death to All Sacred Cows: How Successful Business People Put the Old Rules Out to Pasture. *Fraser is also managing director of the international advertising and corporate identity firm, The Gate Worldwide. I spoke with him about the themes of the book, and what they mean for the radio industry.*

Beau, what is a "sacred cow" in the business world?

A "sacred cow" is a rule, a standard, a formula that we, in business, blindly follow because that's the way things have always been. At one time those rules, those standards, those formulas may have made sense, but unwittingly they became "sacred" over time even though the world, the consumer, the business, the industry has changed. And, unfortunately, a lot of businesses don't recognize that the rules have changed and the world has changed, yet they still use these outdated criteria.

One of your chapters is "Follow the Leader." What makes that a "sacred cow"?

At The Gate, when a client is in trouble they will often look wistfully at their competition who's doing very well (and is invariably No. 1 or No. 2 in the category) and say, "Well, 'Acme Nuts and Bolts' did that, we should do that as well." The problem is that Acme was successful for things that went well beyond whatever strategy or idea they had. That strategy or idea they implemented was successful because of things such as distribution channels or operational issues or brand issues or geography issues. If you can't duplicate those, you certainly will not be successful if you try to borrow their strategies.

Certainly you need to keep an eye on what the competition is doing, because you want to stay abreast with them, but if all you do is duplicate what they're doing, all you're doing is keeping pace.

For example, while a radio station may have the same format from one market to another, that radio station in another market may be successful for things that may not exist in your market. What I always tell our clients is to keep an eye on the competition because you don't want them to get too far ahead of you, but spend more time looking at your own business and figuring out what it is that you do very well and work harder on doing that even better.

If all you do is match your competition's strategies, all you're doing is keeping pace. And in most businesses, keeping pace is not enough.

Okay. Here's another "sacred cow" from your book: "Focus on the numbers, and the rest will take care of itself." Now, in radio there are an awful lot of people focusing on the numbers first. What are we doing wrong?

Let me illustrate this very personally. My responsibility to The Gate is to deliver a certain amount of revenue to my holding company. What they tend to do – and most companies do this – is to say to the general manager, "You need to deliver this much revenue." But that doesn't give really clear directions. To me that's really a goal as opposed to telling people what it is you want them to accomplish.

So suppose instead that I tell everyone inside our organization that our goal is to be agency of the year by 2011. Now, I know that if I achieve that, all the wonderful numbers that investors and employees and owners want will come, but to me numbers are a byproduct of having achieved something else. So I think it's more interesting to think of what it is we want to achieve – for example, a radio station could be "the most favorite radio station in the marketplace," which is more an emotional measure rather than an actual number measure. But if you are the most favorite radio station in a marketplace, all the things that you'd want, such as reach and frequency numbers as well as listenership numbers will come. But these become a byproduct of a larger, more interesting goal.

Now, here's another one, Beau, from your list of "sacred cows." It's called: "Don't screw up."

I like that one because starting off in the business world I certainly screwed up a lot. But the problem, for instance, is if you tell your child, "Don't screw up. Don't screw up. Don't screw up," they're going to spend more time trying to figure out ways to not screw up rather than coming up with a more interesting way to do something.

Look at Pfizer coming out with Viagra, which was not a goal but something discovered by constantly failing on a drug that was conceived to do something else. So very often by making mistakes you can achieve something that is much greater. I'd rather tell my people, "I don't mind if you make a mistake, as long as you can give me a rational reason for what your goal was." As long as you learn from your mistake, I think you're going to come up with far more interesting solutions.

You can't stifle creativity, and I think that, for me, is what the "sacred cows" are all about. They hinder progress, they hinder creativity, they hinder original thinking, and if you block those out, if you eliminate those, you're more likely to succeed – and focusing on making sure you don't screw up is certainly one "sacred cow" that needs to be very gently and humanely put out to pasture.

So then, Beau, what you're saying is that part of creativity itself is screwing up?

That is correct. But I prefer the word imagination. Those who have imagination spend more time thinking about, "What if? How about?" as opposed to using some of the words that hinder.

When I'm talking to a client, and I present an idea and they reply, "We've tried that before and it didn't work," and I ask them why it didn't work, and they can't answer – then I know I'm in the presence of a "sacred cow."

Or the even worse one is someone looking to create an easily repeated formula, because, by definition, a formula always works – you can't argue with a formula – so when someone says, "What's the formula?" I know that I'm in the presence of somebody who holds onto a "sacred cow." And all those things hinder original thinking.

If you think about some of the great breakthroughs in the business world, it was invariably something that was totally unexpected. They had the courage to try it. And it doesn't mean that you have to spend an awful lot of money. If you're a large organization and you try something and it fails, a failure can be quite devastating. You can certainly try it small, and make changes as you go to market.

Okay. Here's another one of those "sacred cows" in your book: "Branding is expensive." Well, branding *is* expensive, isn't it?

I think the assumption is that when people say, "Let's brand something," they automatically think of 60-second commercials on the Super Bowl. But to me, branding is an intellectual exercise, and I don't think intellectual exercises are expensive. Depending on how you execute it, your branding program can certainly be expensive. But even if you don't put any communications behind it, even if you don't put any media behind it, what you should always do is act like a brand. And a brand is what defines you as being different, better than the competition, and gives potential perfect customers reason to choose you. A brand is giving a reason to choose.

To me, the best definition of a brand that I've heard is this: "A brand is what you get when you add differentiating substance to a product." I like that definition because it tells you that you the client, or you the marketer, are in control, and you get to decide how you want to define the brand.

And, to me, the other part of it is differentiating substance. Very often when I read brand descriptions of a product that comes from a company, to me it sounds very blah, blah, blah, or I've heard it before. So it really has to be differentiating, and it has to be substantial – something that people can really sink their teeth into.

Meanwhile all of that is an intellectual exercise and has nothing to do with spending.

To me branding is a noun and not a verb.

What "Positioning" Is

Maybe it's because we have been so indoctrinated in the rules of *Positioning* authors Ries & Trout, the notion that a "position" is a place in the consumer's mind and a "word" is what you need to "own" to occupy that position.

Well a position *is* a place in the mind of the consumer, of course. Because everything everywhere is a place in somebody's mind. This is almost like a nonsense fact.

What's lost in this simplification is that there is a sharp difference between a "position" and the act of "positioning."

Ultimately a "position" is standing for something meaningful to somebody. It is not – I repeat, not – necessarily a "word."

I ask you, what "word" does Starbucks own? How about Nike? How about Hallmark?

But do these brands stand for something? You bet they do. You don't boil it down to a "word" because, just like people, a strong brand's personality has more than one dimension, but all those dimensions fit together in one tidy and meaningful and compelling package.

A "position" is something you "own" not because you claim a word, but because you express your position. In other words, you do who you are, thus you are what you do.

Be who you're trying to be. Don't just say what you're trying to do.

Too many broadcasters are groping for something to stand for, but this doesn't come from a consultant, it doesn't come from research. It comes from the soul of your station's creator and how that soul resonates with the tastes of your audience, your fans.

It comes from inside your brand; it's not stuck to your brand like a Post-It note.

You need to trust your audience to recognize that soul. You need to trust them to recognize your authenticity, assuming you have some. You need to trust them to understand your "position." All you need to do is express it.

What do I mean by "express your position"?

I want to illustrate by example. The comic book on the following pages really is a comic book. Or, rather, a comic book that doubled as a direct marketing piece to thousands of radio stations by my company in 2007.

I could have sent a postcard saying "the best research" or "tops in branding" or some such puffy nonsense that will land in your circular file long before it lodges in your brain.

So I didn't.

Instead I created this comic book.

But it's not pure entertainment. Indeed, there's a message embedded in this visual spoonful of sugar.

More than one broadcaster told me "I don't know if people will get it."

But I have more trust in my audience than that.

mercury
radio's perceptual specialist™
presents
Radio
NOIR
MARK
RAMSEY
DANY
BOOM

WHEN THE SAINTS ARE SLEEPING IN THE CITY OF ANGELS, THE STREETS ARE LEFT TO THE SINNERS....YOU CAN SQUINT AND SEE THE MOTHS SLAM THEMSELVES AGAINST THE GASLAMPS.
SMAK!
KLUD
THERE'S A LOT OF SLAMMING GOING ON, AND IT AIN'T ALL MOTHS.

FOR A SAWBUCK YOU CAN BUY ANY KIND OF TROUBLE YOU WANT.
mercury
INVESTIGATIONS
BUT WHEN THE TROUBLE COMES HOME, THAT'S WHEN YOU KNOCK ON MY DOOR.
THE NAME'S MERCURY, AND I'M IN THE INVESTIGATION BUSINESS. I HELP CLIENTS SOLVE PROBLEMS.

THERE ARE SOME NIGHTS I SHOULD FOLD UP MY TENT AND CLOSE EARLY
AND TONIGHT WAS ONE.
KNOCK KNOCK !
"yeah ?"

"Mr.Mercury ?"
mercury
"Who's asking?"
"I need your help."
"Everybody needs something, sister."
mercury
"My husband has been murdered."

mercury
FOR A DAME WITH A DEATH IN THE FAMILY, THIS DOLL WAS GIVING ME A LOOK SO DIRTY I COULD USE A COLD SHOWER...
"I ain't interested, baby. Call a copper."
HER HIPS WERE SWAYING AROUND MY OFFICE -
I WAS GETTING DIZZY JUST TRYING TO IGNORE THEM.

"The law can't help me. I want you."
"And I can pay you."
HER HAIR WAS CASCADING DOWN HER SHOULDERS LIKE NIAGARA FALLS - THE CANADIAN SIDE. I GULPED HARD ENOUGH TO SWALLOW A BOWLING BALL.
THIS WAS THE KIND OF DAME WHO GETS INTO YOUR BLOOD LIKE HEPATITIS C.
"I want you to help me solve my problem. His name is Horace Delacorte, and he murdered my husband."

HORACE DELACORTE. KING OF THE SCUM. ONE-TIME CARDSHARK TURNED CON MAN TURNED CONTRACT KILLER TURNED MOB BOSS. NOT ONE OF ANGELINO'S FIRST CITIZENS. AND A MAN THIS GUMSHOE HAD LONG SINCE COME TO LOATHE.
"Will you help me?"
"You've got a chip on your shoulder, baby."
"That's a love bite."
"Sure, I'll help you. Starting tonight."
"Swell."

"It's cold out, you'd better wrap up in something warm."
"Is that an invitation?"
"Every night is an invitation, Mr. Mercury. Good evening"
Mercury
I WAS ON MY WAY MINUTES LATER. I KNEW WHERE TO FIND DELACORTE, AND I WASN'T WASTING ANY TIME.
AND I WASN'T ALONE.

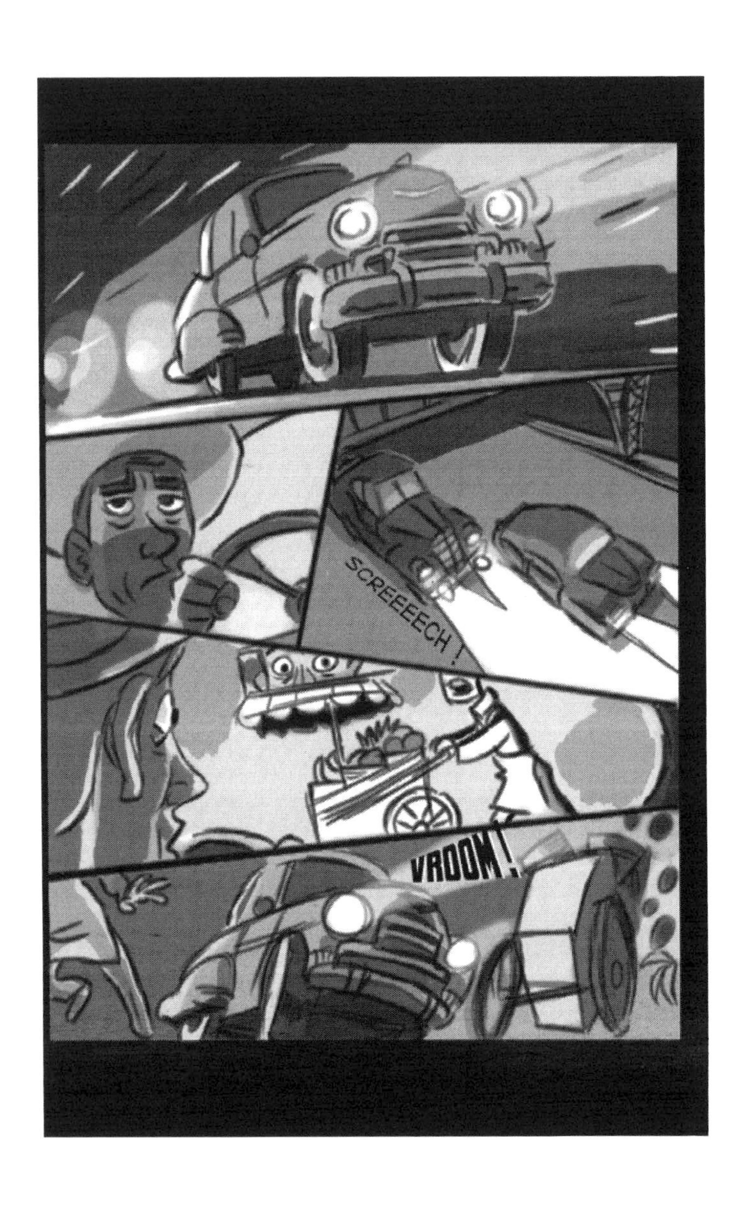
SCREEEEECH!
VROOM!

"How could I hit a fruit cart?"
"Somebody always hits a fruit cart."
SMACK !
"Mr. Delacorte understands you have a new client."
"Mr. Delacorte wants you to say thanks but no thanks to your new client. You're in over your head."
"Say goodnight, Gracie."
CLICK !

"Next time I'll remember the bullets."
I HATE DOCTORS. I HATE HOSPITALS. I HATE SPENDING TWO WEEKS WITH MY ARM IN A SLING.
BUT THESE GOONS HAD ANOTHER THING COMING IF THEY THINK I'D GIVE UP THIS EASY.
SO IMAGINE THEIR SURPRISE WHEN I PAID THEM AND THEIR BOSS A LITTLE VISIT AT THAT GIN JOINT DIVE THEY CALL A CLUB.
"I'm here to see Delacorte."

"Come right in, Mr. Delacorte is expecting you."
"You've got balls, Mercury. But not for long."
"I don't take kindly to having my clients hassled, Delacorte. Nor do I savor being mashed up against what used to be my car."
"You think you own this town Delacorte. But you're just a dirty bum and you don't own a hill of beans."

"You put the squeeze on my client, so I'm here to put the squeeze on you."
"Big talk from a palooka who's about to get clipped!"
MAYBE THIS WAS WHEN IT OCCURRED TO DELACORTE THAT I HAD MY EYE ON HIM FOR WEEKS.
MAYBE THIS IS WHEN IT DAWNED ON HIM THAT I KNEW HE CAME TO THIS CLUB EVERY NIGHT, THAT I KNEW WHERE HIS TABLE WAS, THAT I KNEW WHAT SEAT HE ALWAYS CHOSE.
THAT I KNEW THE SEAT ACROSS FROM HIM - THE ONE I WAS IN - WAS ALWAYS EMPTY.
MAYBE THIS IS WHEN HE REALIZED I COULD SLIP IN BEFORE HE ARRIVED AND HIDE A SURPRISE WHEN AND WHERE I'D NEED IT.
MAYBE THAT'S WHY I SAW TINY BEADS OF SWEAT ON HIS FOREHEAD.

TAK!
TOK!
PAK!
"Put it on my tab."

IT WAS A BUSY WEEK FOR THE TRASH COLLECTOR, BUT NEITHER THE COPPERS NOR THE GOOD PEOPLE OF LOS ANGELES ARE GRIPING.
I HAD A CLIENT, AND I SOLVED HER PROBLEM. 'NUFF SAID.
GANGLAND TRIPLE SLAYING
BUT WHAT ABOUT THE DOLL? I NEVER DID SEE HER AGAIN.
GANGLAND TRIPLE SLAYING
MAYBE I'LL FORGET HER...MAYBE I'LL DIE TRYING.

Being Authentic in a Paris Hilton World

An interview with writer Bill Breen.

Bill Breen is senior projects editor for Fast Company *magazine and the author of a piece in the May 2007 issue titled,* Who Do You Love? The Appeal and Risks of Authenticity.

Bill, who cares about "authenticity"? What happened to just being good?

People are always calculating, "Is this thing real or not?" but it almost happens subconsciously. Consumers have gotten very sophisticated about marketing and can instantly smell when someone is not sincere, when they're not real, when they're not acting with integrity.

But I think the hunger for authenticity also comes from our sense of disconnect from what is really "original" in life. The source word for authenticity goes back to the Greek word for originality – there's a search underway; it's almost a spiritual thing.

When you start really digging at this notion of authenticity, you find that we're also willing to accept a version of "fake real" that can become more "real" to us than the real world. Think about the success of Jon Stewart. He cuts through the pomposity of network news. And in that sense, he is real. He is authentic, even though what he's giving us is fake. And people respond to that.

Authenticity can be created, and that's part of what the article is about.

So you're saying that authenticity isn't about being real per se. It's about not being fake.

Well that's part of it. I have a line in the piece which says, "The opposite of authenticity isn't fake, it's cynicism." Think about the cornerstones of authenticity: One is "integrity" – you are who you say you are.

Now consider an insincere marketing campaign like this one from McDonald's – "We love to see you smile." It was an abject failure because consumers really didn't believe that McDonalds made them smile. Even Ronald McDonald was a clown who didn't make you laugh.

And so the message did not connect with the reality of the experience. And that's one of the core things about authenticity: What you're saying has to be real, and the experience and the communication around it have to connect. If what you're saying doesn't sync up with what you're actually producing, you will breed cynicism in people. And that's the danger.

When you look at Starbucks, it is built around a fabricated experience of Milan espresso bars. And in its early days it pulled that off beautifully, and it was very successful doing so. But now a lot of the things that made Starbucks feel authentic have been diluted as the brand has grown and grown and grown. And when you lose authenticity, you end up in a world of trouble. Levi's has gone through the same thing – it's an iconic American brand, but it didn't stay relevant to its times. And that's one of the hard things with authenticity. You have to respect your values and your heritage and yet evolve at the same time or else you become irrelevant.

Your challenge is to stay true to your values but not get bound by them.

What makes Apple so much more "authentic" than its numerous competitors?

I think one of the cornerstones of authenticity is the idea of serving a larger purpose. And Steve Jobs really does believe that he and his company are there to do that. His mission is to change the world through technology and design. That's something consumers can get behind – when you have a mission, when you think you're serving a larger purpose.

Every business has a "money story," but a business needs to be about more than that "money story" in order to be "authentic" and win the affection of its market. Apple comes off very well in that regard, and Microsoft does very poorly. Microsoft is the ubiquitous company that's largely unloved because people get the sense that it's only about the bottom line there, that they don't particularly serve a higher purpose despite some of their marketing claims.

So to be authentic you need integrity and to serve a larger purpose. Your piece also mentions the importance of a strong point of view.

Right. I live in the Boston area, and I listen to WEEI, a sports/talk station. I constantly find myself making the calculation, "Are these guys authentic or not?" And I think they score very highly on the passion point of view. I mean this is "Red Sox Nation" out here, and they are very, very passionate for the team, even hypercritical. And they also exude this sense of place which I think is another criterion – authenticity springs out of the idea that there's a place here, a place with a story. And these guys know Red Sox baseball or any New England sport inside and out.

But at the same time (and I find this happens a lot with AM stations), there's the marketing side, the hyper-masculine voice you hear on the promos: "We've got the best sports talk in town, and don't listen to the rest of those weenies." Or promos like, "Be the fifth caller and you'll win a couple of tickets." That's when listeners sense that their message, their passion, is not syncing up with their communications.

Look, I'm willing to accept the ads, but when the promos start getting in the way I find myself turning the channel to something else. That's one example of how we make these subconscious calculations. "Are these guys being authentic at the moment? Am I really gonna stick with them?" And if they are, we stay. If they're not, we're outta there.

Too often, radio promos are clichés. And clichés by their very nature, by their very definition, are insincere. There's nothing authentic about them.

What's real is these guys when they're actually talking about sports and their passion comes through. Then it's enjoyable and fun and entertaining and all the rest.

So where's the line between positioning language that communicates with consistency and inauthentic and damaging clichés?

Words are powerful. But marketers in general and radio in particular must stop thinking about their audience as a market and start thinking about it as a community.

When WEEI's hosts are talking about sports they have a sense of place. They're passionate about it. They have a community of people who care about what they're talking about. They are community-builders.

But when the promos come on, that's the marketing side. Now I'm getting the "money story." And if it becomes too much about the "market" and not enough about the community, listeners will shut it down.

Making Radio Hilarious

An interview with actor and writer Ricky Gervais.

Can a podcast be as popular as a radio show? You bet it can – if it's hosted by Ricky Gervais. Ricky has roots in radio – he was once a fixture on radio in the U.K. But he's better known as the co-creator and star of the original BBC version of TV's The Office. *He went on to co-create and star in HBO's* Extras, *terrific shows, both. He's the winner of two Golden Globe Awards. And he created one of the most popular podcasts in the short history of the medium.*

You've won a Golden Globe – just like Sarah Jessica Parker, Rick!

Well, I've got two, though. How many's she got?

I think she's got about 39.

(Laughter)

Exactly, yeah.

You're also the man responsible for introducing Kate Winslet to foul language. Rick, did she know all those words?

No, she didn't. We had to spell them out phonetically for her, yeah.

Now you host the *Ricky Gervais Show*, which is a podcast – one that consistently ranks in the top five at iTunes. And as you like to say, it's very often number one. Why in the world did you do a podcast?

Because I do think I enjoy everything I do now and always have done. I've done it for one reason and one reason only: The joy I get from the creative process.

It's – it's absolutely non-restrictive, there's no one to tell us what we can and can't do. We can do it as long as we like, we can stop if we like, we don't charge for it; it's free.

But it's the favorite thing I do at the moment, you know? To be in a room, shouting to friends, talking nonsense, and releasing that on the world. I made it sound like anthrax. It's just such a joy.

It's so funny because we went straight to number one. I think we now regularly download about a quarter of a million a week. We've done over 2 million downloads since it started.

And the first thing that [Ricky's podcast and writing/producing partner] Steven Merchant said was, "Why aren't we charging for it?"

(Laughter)

And we may have to because it's costing us money, thanks to all the downloads. We're getting so many that it's actually costing money.

(Laughter)

So we might have to charge like 50¢ or something to keep going. But at the moment it's free, and it's just so much fun and it's just a joy. It's me, Steven Merchant, and Karl Pilkington, who is the closest thing to the missing link you'll ever meet.

Well, let's talk about him.

I mean it's like they shaved a chimpanzee. It's incredible. But it's so much fun.

One of the main themes of your show, by the way, is the spherical nature of Karl's head, which really is uncanny.

It's mad, isn't it? I mean it really is like an orange.

Karl's head makes him look like E.T., and I'm thinking that there's an off-chance that Karl could indeed be the next evolution of our species.

You see that would be frightening if it was going that way.

I would assume that you've missed a few million years. But you might be right.

It might be right – maybe – I mean he's happy in himself. He wanders through life – he looks like Charlie Brown grew up and kept the same amount of hair.

True or false, the real star of the show is Karl Pilkington, your producer?

There's no doubt about it. But we knew that. I did this to try and make Karl Pilkington famous.

I consider myself rather like the guy in *The Elephant Man* who takes around John Merrick. I want to take Karl to medical seminars. I want to take him around museums and hospitals and say, "Look at this! Look at the roundness of his head!"

(Laughter)

Karl's favorite film, by the way, is *The Elephant Man*, and I was saying, "Why is it your favorite film?" and he went, "Well, because you know what you're going to get."

Because you know what you're going to get!

That's it!

(Laughter)

And then there's that bit where Anthony Hopkins takes him behind the screen and reveals the Elephant Man, John Merrick, naked, and he's talking to surgeons and he's going to see this disfigurement of the skull – see the spine – see the little – see the – ironically, the only thing that is totally normal is his genitals. Okay?

So Karl Pilkington says, "Oh, think about it – the one thing you would want like an elephant...."

(Laughter)

We'd love to do an "Eliza Doolittle" on Karl. Just see if we can sneak him into regular society without him saying something stupid.

So does Karl know he's the star of the show?

Uh, yes. Yeah. He knows he's got the same celebrity status as something in a jar.

(Laughter)

Now, you say you did these podcasts to enjoy the creative process. A lot of radio people would like to tap into your secret – how do you create such an incredibly compelling show?

They're all shouting into the radio now – what does he mean "creative process" – there's nothing creative about it at all. It's a rambling, friggin' madman!

So what is your creative process? I mean, how much of this do you sit down and write? How do you prepare for it?

None of it. None of it. None of it at all.

We meet beforehand. We look at some emails.

There's always a couple of emails that strike us. We get about 4,000, so, you know, we're trying to get through them and we pull out an email that will start a conversation.

I might have seen something on the TV that I save for Karl and we bring it up. I mean it's all about what Karl says, really.

Steve and I can start the conversation, but we know that if we're talking about anything, Karl is going to open his mouth and something remarkable is going to come out.

I mean it's just – it's as simple as that, you know? We were talking about Bill Gates – some statistician had worked out that Bill Gates could give everyone on the planet $6, okay?

Uh-huh.

Karl said, "But that's not fair, is it?"

And I'm like, "What do you mean?"

He said, "Well, a little fellow living in Africa, he doesn't need $6, right?"

And I went, “So you’re saying he should give more to the *rich* people?”

And Karl went, “Well, yes – $6 doesn’t go anywhere in New York, but in Africa give him $2 and he’s happy.”

I mean this is the remarkable brain of Karl Pilkington. He sees the world in a different way from everyone else.

(Laughter)

When you try to create something funny – entertaining – and whether or not you plan it in advance, what do you think the essential ingredients are?

Well, it’s context. It has to be contextual. You have to let people know what they’re watching and why, and they have to have a reason.

I think warmth is important. Whatever it is – the comedy can be quite harsh and quite out there – but there has to be a warmth, there has to be a warm vessel. You have to like the person you’re not meant to like.

You have to – do you know what I mean by that?

Yes.

For example, Karl can be saying ridiculous things, but you can’t dislike him because he’s childlike. I mean telling Karl off for something he said is like rubbing a cat’s nose in its own wee. It doesn’t really know why you’re doing it.

Only it’s a little bit crueler; then you feel sorry for it afterwards. So there’s no point.

Also, I think you have to not pander, not try and please everyone.

I think there’s no one rule; there are lots of rules and then again you can ignore them all. So I haven’t said anything have I?

No, I think you’ve said a lot.

I’ve not answered the question or helped anyone.

I think you’ve said quite a lot and answered the question completely.

I think I’ve said too much.

Now, in your writing process, whether it's TV or radio, how do you write funny?

Well first of all you don't write funny, you *think* funny and – and then the difficult thing is writing it.

And, you know, I think that's a metaphor for TV comedy. People can come up with funny things, they can write funny things, they can be funny, but they can ruin it in the rendering of it on TV.

They can really miss out there and often that's the middleman's fault. And that's why my favorite stuff is auteured and I knew that what I had to do was cut out the middleman. I knew I wasn't the best writer in the world, or director, or actor, but I knew what I wanted it to look like on TV.

And that's the most important thing.

I think it was Woody Allen who said that the best an idea gets is when it's still an idea and it's just a matter of how much you ruin it between your brain and the TV.

And that's what you're fighting against, not letting anyone ruin your idea. And so the answer is do it yourself or stop whining.

Is that why the content you create is so tangibly different from so many of the sitcoms that fill the network schedule?

Yeah, but I didn't do it because I thought those sitcoms were bad. I just thought they weren't "me," and why do we need another one?

You know some of my favorite shows are, you know, live studio audience. *Seinfeld* was a live studio audience but it made sense, it made perfect sense.

You know, I didn't want to completely get rid of the idea of a central character. Some of my favorite shows have central characters –*I Love Lucy*, *Sgt. Bilko*.

I didn't want to lose the wit. Some of my favorite shows are witty, like *Cheers* and *Friends*. But what I did want to do is change everything a little bit.

It's as simple as that. You know? So, it wouldn't make sense for a fake documentary to be watched by a studio audience, so that had to go – that was a necessity.

Because *The Office* was a fake documentary and the realism was, you know, absolutely central.

Most people aren't ceaselessly funny. Most people aren't Chandler. Most people are awkward. And so we couldn't have too many Chandlers. You know, Tim [renamed "Jim" in the American version] was the only "Chandler" that we could have, and he didn't get the laughs.

Instead, the people saying stupid things that *weren't* funny got the laughs.

We didn't want convoluted plots because in the normal office, not much happens. So we had to contrive that a little bit and make it look like it was cobbled together. So all these things came from, I suppose, taste.

And realism. In fact, everything came from that.

Of course we could have gone the completely wrong way. It could have been the most realistic, wonderful document of life in an office and not been funny or interesting. Because we might have forgotten to put something in that was interesting, so we had to really walk that knife edge.

You have to try your hardest. Too many people do a bad show because they've run out of ideas – but the money's still coming in.

I think you've got to know what you set out to do, and when you've done it, walk away. Not be greedy, not do anything for the money, not pander, not worry about ratings, not worry about popularity or awards.

And if you do that, all those things will come.

In your process, is it more effective for you to try to be true – to be real – or to be funny?

Well, my *job* is to be funny. My *passion* is being true and honest.

And, again, that's the line I walk. I really want to do both. People tuned in for Brent [Michael in the American version] and they stayed watching for Tim and Dawn [Jim and Pam in the U.S.].

So you need both.

Now you're a producer of the NBC version of your show. When you heard about the early testing on that show, that it was supposedly "the worst testing TV pilot of all time," what did you think?

Yeah, the moods of the American producers were all down, and I sent [producers] Greg Daniels and Ben Silverman an email where I said, "All bad news is it? That's *great!* That's exactly what happened in the U.K."

And it's absolutely true.

It scored the lowest ratings ever on BBC-2.

And so I said, "That's great boys, that's a great omen. Keep up the good work!"

And right you were; it's performing much better now, isn't it?

Yes.

Don't pander.

Don't sell out.

Don't worry about what you should be doing or who's going to watch it. Do what's in your heart and do what's true to your thing.

There are 6 billion people in the world; there's going to be enough of them that'll weigh in for something you created and it's going to be enough.

That's what I meant, when it really matters. When you didn't sell out and you made it. Because if you sell out you could still fail, and if there's one thing worse than selling out and making it, it's selling out and failing.

So the answer is don't sell out.

One last question: How long are you going to let Steve Merchant hold you back?

(Laughter)

As long as he does my laundry he gets 20% of everything I do.

Steve saw that actors get paid more than writers and he said, "Right, Rick, I'm being on *Extras*." I went, "Yeah, okay."

And we did our scene together [On *Extras*, Gervais played a struggling actor and Merchant played his agent] and I started laughing at a face he pulled and I went, "Why'd you do that face?"

And he said, "I wasn't pulling a face."

That's got to be worrying hasn't it?

And I said, "You look like Beaker from the *Muppets*."

And he went, "That's what I was going for."

That's his aspiration!

I try and be like Al Pacino or Robert De Niro – he aims for Beaker from the *Muppets*.

And you know what? I think he gets that.

Seth Godin on Radio's Future

An interview with author, blogger extraordinaire, and marketing guru Seth Godin.

Seth Godin is the author of many terrific marketing books, including The Dip, Meatball Sundae, Purple Cow, *and the classic* Permission Marketing. *The author of one of the most popular blogs on the web, http://sethgodin.typepad.com/, Seth is an icon among marketers, and this interview is easily the most popular one I ever did with any marketing guru about radio.*

Seth, there's a lot of discussion in media circles about *The Long Tail*, but your book *The Dip* specifically focused on the inverse of *The Long Tail*, namely, "The Big Head." Can you talk about what that is and the significance of it?

Sure. Chris Anderson wrote a book called *The Long Tail* that basically explains that if you give people lots and lots of choices, many of them will take those choices; so for example, when Amazon offers millions of books for sale, they will sell more books than they would if they only had a few books for sale, which seems like common sense.

What's real interesting, though, is that Amazon.com gets half its revenue from books that the brick and mortar stores don't even carry. What's interesting is that if you add up all the Internet radio stations in the world, half of them are playing songs that are never, ever, ever on traditional radio. If you add up what gets sold on iTunes, half of all iTunes sales are of songs that aren't carried in a record store. So the lesson of *The Long Tail* is that way out at the end where the obscure records are, there is still money to be made.

But, if you have to choose between having one title out at the end of the tail or one title up in the short, fat, juicy head, as I call it, you should pick the one that's up at the top. As the "Long Tail" stretches out, the hits are worth more than they were ever worth before – what is paying off these days is being a superstar.

So Google gets way more searches than Ask.com because it's the easy one to pick, and Paris gets way more tourists than Tampa because it's the easy one to pick. So once something becomes a superstar, once it gets to be in the scarce collection of winners, then the Internet drives way more people to it and it continues to succeed. And so the lesson of the book, I think, is unless you're lucky enough to be Amazon or the iTunes store or someone who can own the whole tail, you're better off figuring out how to break through the clutter and be in that short head.

Then why have a "Long Tail" at all? Why not just create hits?

The answer is surprising. The answer is that most of the things we set out to do in our lives are controlled by one of two curves. Most things are dead ends or cul-de-sacs. They are flat paths. They're hamsters on a wheel – somebody who is doing the same thing every day pushing along, pushing along, but never breaking through.

But a few things are controlled by "the dip," and the dip is the screen, the filter, the thing that separates those who are scarce, the professionals, the superstars, from the masses, from the amateurs.

So, let me use a couple examples from the music business to make this clear. In 1968, the dip was getting a record contract. The dip was whether Clive Davis picked you out of a crowd, because if you made it through that, then you were a professional. Then you had a shot. Then you were going to go on tour and sell records, and if you didn't, you were a failure. You could see that dip and you could identify it and work your way and maybe get through it.

Well, today, that dip has disappeared. And if you're an independent musician today, the good news is you've got a chance at making it without that. The bad news is without that dip, without something to get through, it's a lot harder to keep slogging. It's a lot harder to know that there's going to be scarcity on the other side.

Here's another example: Suppose you're a standup comic, *The Tonight Show* is your dip. Either you've been on the *The Tonight Show* or you haven't, and until you have, you're not going to make it on the road because that's the thing that separates. It's like taking organic chemistry on your way to becoming a doctor. It creates scarcity.

And so, what we need to do as organizations and as individuals is to seek out dips, not avoid them, but seek them out. Find those hard things that if we just focused all our energy and talent on, we could get through, and if we can't find one, if we're on a dead end, we should quit what we're doing immediately and go find one.

So contrary to the Donald Trump advice of never, ever, ever give up, you're saying, in some cases, you should absolutely give up because you're faced with a cul-de-sac and you have nowhere to go.

That's right. Take the Space Shuttle. Our government spends more than $1 billion launching each one, 50,000 different people work on each launch, but still the Space Shuttle is never going to get better or safer or cheaper. It's broken. We shouldn't have it. We should quit right now. We should stop the Space Shuttle program, because if we did that, we would create an emergency, and the emergency would cause lots of talented people to put enormous amounts of effort into making something significantly better than the Shuttle and we'd fix the problem.

But as long as we say, "Well, we're doing our best under the circumstances, and we're trying really hard," we're always going to have a mediocre product.

Put yourself in the position of a radio broadcaster. Their aspiration is for their station to be number one and they're trying, trying, trying, trying. How do they know that what they're in the midst of is a dip or a cul-de-sac?

If you are running a radio station with the consultants and the conventional wisdom and trying as hard as you can, you need to ask yourself an honest question, which is: Is it likely to ever be any better than it is now? Meaning, is satellite radio going to become less popular? Is Internet radio going to become less popular? Are people going to find fewer things to do when they're in their cars?

I think the answer to all of those questions has to be "no," that traditional, terrestrial radio is a zero-sum game. In fact, it's worse than a zero-sum game. It is clearly headed towards a dead end. There's no dip ahead. There's no breakthrough that's going to occur.

But, you have all these assets. You have advertisers. You have access to creators of content like record companies. You have access to some listeners. Why not use those assets, use that leverage to build something new where there may very well be a dip. If I ran a radio station today, I would say, "How do I get every one of my listeners to sign up so I can have a direct relationship with them by phone and by email? How do I learn what their zip code is? How do I discover what they're interested in?"

Because if I could do all those things using the assets I have now, I could find a new dip and get through it. I could be the go-to source for where listeners should go when they want to party, where they should go when they want to go out for dinner, where they should go when they want to buy a car. And if I use the stepping stone of my terrestrial FCC license to create something new, knowing that I'm going to quit radio as soon as it's done what I need it to do, then I can move forward, find a dip, and build behind me a moat, a valley of death, a dip so deep my competition could never get through it.

And the station's Internet strategy is the key to transforming listeners into relationships?

Yes. I want you to close your eyes for a minute and think about what the world will be like when every car sold has Wi-Fi in it – and it's only four years away. When Wi-Fi blankets the city of Philadelphia and every car has a Wi-Fi receiver, that means I can listen to a million channels, not 10. And what's that world going to be like? Every listener is going to choose the one that's tailored for him or her and it won't necessarily be based on geography.

But when I look at a typical radio station website, what I see is every radio station trying to do a slightly above-average job of building a radio station website. I don't see any radio station saying, "How do I completely reinvent my interactions with users so that when my FCC license is worthless, I'll be glad I did what I did?"

So what advice would you give to the management that has to answer to the stockholders and the Wall Street folks to say, "Look gang, we've got to get through this dip. It's going to be time consuming. It's going to be expensive. It's going to be worth it." What do you say to them when they're face to face with the money guys?

Well, Wall Street has doomed a lot of companies. Wall Street doomed AOL. AOL saw the web coming. If they told Wall Street the truth, AOL would be Google today. And it isn't because Wall Street scared them.

Now it's not your fault, Mr. Radio Station Man, that your company went public. But, you are public and now you have two choices. You can either say, "Wall Street's going to force us down a dead end, I'd better make sure my pension is fully funded," or you can just go to Wall Street and say, "Look, this industry is changing. We have a plan. It's going to take several years to get there. We're going to continue running radio stations the best we can. But, guess what we're going to build for the future? For the future, what we're going to build is space-based, location-based mobile interaction on a custom basis that no one will ever be able to surpass, because we're not going to define ourselves by our FCC license anymore. We're going to define ourselves by how many people have come to us and said, 'Here's my contact info. Here's my Twitter address. Here's my Facebook info. I want you, Mr. Information Man, to keep me up to date with music and information that's geography-based for the rest of my life.'"

And if you don't start doing that now, there's zero chance you're going to be able to do it in five years.

You're talking about leveraging the strength that stations have with their audiences, the strength they have with their advertisers?

Right. The beauty of this is that the guy who runs the Chinese restaurant 40 blocks away, he doesn't need to advertise on Google and he never will. He's not going to advertise on television or YouTube because it doesn't make sense, but he advertises on radio right now, and he advertises on radio on a local station because it pays for itself. And geography is what radio stations have going for them; they are local.

And so, if each radio station figured out how to find the 12,000 people who that Chinese restaurant and that laundromat and that car dealership need to reach regularly, and could spend all their time not yelling at them with ads, but doing that local interaction, that information and entertainment that these people want, then they're going to have an asset that's worth $100, $200, $500 per person. Add up the math. If you own enough zip codes, you win.

Now, technologically speaking, AM stereo was once thought to be a dip but it turned out to be a cul-de-sac. One of the solutions that broadcasters have proposed to combat the 10-million-station Wi-Fi universe is to create more versions of stations via HD radio, so every station becomes two or three stations. Is that, because it's an extension of the license, a non-solution to the problem, in your view?

Well, what *The Long Tail* shows us is that once you start spoiling people a little, they want to be spoiled a lot. Starbucks used to have 100 combinations of beverages and now they have 19,000. When I turn on the radio, I want the traffic report only for people who are leaving my neighborhood and going to where I'm going in New York City. I don't care about the Kosciusko Bridge. I don't even know where it is. Don't talk to me about it.

And you know what? I've heard that Elton John song too many times; I never want to hear it again, and on and on and on. I want it to be about me all the time. I want to hear the report about my school district, not a school district that has nothing to do with me, and I don't like basketball, but please, please tell me about what happened at the billiards tournament. And there's all this stuff that I want that is going to get customized for me by someone who isn't hampered by a history of owning radio stations, and so if you're going to walk in with this radio station thing on your back and say, "No, we're only going to have four flavors because the technology won't let us have 80," then I'm just going to leave because I can.

But under that scenario, where is the space for a hit when everyone wants their own hits?

I'm so glad you asked that. The book talks a lot about a phrase I call "best in the world." If you're the best in the world, people wait in line for you, people call you on the phone, people ask for you. But we have to talk about what "best" means and what "world" means.

"World" doesn't mean the whole universe. It doesn't mean the planet Earth. It means "my world." Best in my world means for what I want right now, are you the best? And "best" doesn't mean most expensive or even highest quality. It just means the one that most matches what I need right now.

So, for someone in the United States who wants an obscure book, Amazon is the best in the world at selling it to them. Between the free shipping, the fact they know who I am and everything else, it's easy to pick them. That's why their sales online are 10 or 20 times their nearest competitor.

But, if you're talking about bread, then for me, best in the world means that guy who makes whole wheat bread six blocks from my house, because for me, I'm not willing to travel more than 20 blocks to buy bread. And best for me doesn't mean the fluffiest or the cheapest. It means whole wheat.

So, best in the world, when I'm driving from my office in Irvington to mid-town Manhattan, my world is that universe, and so there isn't going to be one radio station that's the best in the world. There's going to be thousands of them for different people, and the beauty of people, when they're approaching it from the radio station point of view, is if you realize that you could own 1,000 stations or 10,000 stations, each one of them could be the best in somebody's world.

So owning 10,000 radio stations is one strategy?

It is. The other strategy is to have content that's so compelling that large numbers of people tune in because it's a hit. It's the Fergie strategy. Fergie's a hit because...she's a hit. People are listening to Fergie because she's on the radio a lot. She's on the radio a lot because she's on the radio a lot, and the fickle finger of fate points to a Fergie every once in a while, and it's great work if you can get it. So, the number one way to become a millionaire is to find $1 million on the street, because it's easy and it works, but I don't think it's a plan.

But it could be a plan if you're talking about non-music content. Then you're talking about something that is even scarcer than music, isn't it? So, if you're talking about sports content, the Howard Sterns of the world and so on?

Right, but the problem with this strategy is you have to predict it in advance, or you have to pay full retail if you come in later, so if you want to put the Yankees on, George Steinbrenner's not going to give you a discount because you're a nice guy. He's going to charge you what it's worth. And so, in my case, my blog is one of the 20 most read blogs in the world, and I'm really thrilled and lucky, and I know I'm lucky. It didn't happen because I deserved it. It happened because it had to be somebody and it was me, and if I told you when I started writing my blog five years ago that my strategy was to have the most popular business blog in the world, you would be right to laugh at me, because it was a total crapshoot. That's not a strategy. That's just hoping.

But if someone came to me and said, "We want to pay you to syndicate your content somewhere," I'm not going to say, "Well, it was luck so you can have my content for nothing." And so, my point is, there's this shuffling of the deck that's going around, the biggest shuffling since the 1930s. And you need to make a decision, which is are you going to wait for someone to get lucky and then pay them a lot of money to partner with them, or are you going to set yourself up with a strategy that's based not just on luck, but on following these paths that people before you have followed in other industries and it's working?

You know, when Google set out to challenge Yahoo!, they didn't say, "How do we out-Yahoo! Yahoo!?" They said, "We're going to do one thing: Search. And we're going to tell people that's what we're the best at, and if you want search, that's what we do and that's all we do. And by establishing that, they pushed through a dip that other people were too distracted to even pay attention to. At the time that they did that, Yahoo! had 185 links on their homepage and Google had two. And so, Google could obsess about one thing and put all their resources and all their focus on doing that, and I don't think it's luck that Google accomplished what they accomplished. I think it's a result of focus and betting everything on pushing through a dip that, at the time, wasn't so big, but now their competition could never surmount.

So, from the radio perspective, you might be saying, "Take a breath, stand back. It's not about what's hot now. It's not about what Clear Channel is doing in market X or CBS in market Y. Take a breath. Stand back and think through the way these trends are going. Think through the potential for your brand and your capabilities in the future, and think it through from a strategic standpoint and don't just plug into whatever the hot trend is."

Well, every time I turn off a radio station, because I get out of the car, because I'm done with something, it may very well be the very last time I listen to that station, because we know that every day, some people decide never to listen to the radio again because they're going to replace it with something else. And the question I would ask someone who runs a radio station is, how are you going to get me back? Because if I turn you off, you don't know who I am. You can't come back and get me. I'm invisible.

And so, every day that you're talking to me where I'm choosing to listen is your chance to build a relationship with me, a real one. Not one where I can sing your jingle, but one where you know my email address and I want to hear from you, where you're a Facebook friend of mine, where you're putting something in my RSS reader every morning on my way to work. And if you can build those hooks and create those relationships, then you have a chance, when I start my new media habits – and they're coming – to be on the list.

And so, what would I do if I ran a radio station? I'd focus on one thing, say RSS feeds or email subscribers, and I'd say, "Tell me where you live. Every morning the traffic report will be waiting for you when you wake up." Now, it's easy for an old-time radio person to say, "That has nothing to do with radio," and I would say that's exactly right. What it has to do with is building relationships for the long haul with people who want to hear from you, and that's the opportunity that every radio station has today that's an advantage over every entrepreneur that doesn't.

I started a website a couple years ago called Squidoo.com. And what Squidoo.com does is to help anybody build a webpage about anything they're passionate about, and so it could include links to books, to radio stations, to concert tickets, to movies, to anything you want to describe as your take on a topic. And when we started, we obviously had no traffic, and I went to, I think, the most popular FM radio station in New York – certainly one of the most popular – and I said, "We could work together, and bit by bit, you could build an audience, your listeners could create their own pages here, and if every one of those pages linked back to you, you're going to have a presence in front of a lot of people."

And because of the politics of the way these things go, and it's totally understandable, nothing happened. What's interesting is that a year and a half later, our traffic is 25 times bigger than their traffic. And I'm not saying that to show off. I'm saying that it wouldn't have cost them anything to use unsold airtime to get all their listeners to create these pages, and all these pages would have kept pointing back to the station, which would have pushed them up. But the mindset is either, "We're in the radio business," or, "We're in the relationship business," and you have to pick one or the other.

Let's shift gears to the music industry. How screwed is the business of music?

Clearly, music is more widespread and more easily available than ever in history, so music and songs and musicians aren't going away. It has been demonstrated that human beings like it. It's not going away. But, the business of it, the business of fairly anonymous labels, labels that have no brand, labels that have no connection directly with listeners, extracting 80, 90, 100% of the profit from the musician, that is clearly going away. And it's going away because polycarbonate and vinyl made it possible for a record to cost $10 or $15, and when you're charging $10 or $15 because of your cost of goods, there's plenty of room left over for the musician and for the label.

But when the cost of delivering music is zero, there's not a lot of room left for anybody to extract money from selling the song itself. And treating your listeners like criminals and suing them hasn't worked, and it won't work, because either they're going to avoid you or they're going to ignore you or they're going to go to jail – none of which are good outcomes.

What will work is realizing that spreading music spreads ideas, and once people like ideas, they like to buy souvenirs. And souvenirs – concert tickets, t-shirts, kinds of interactions that are not scalable, kinds of interactions that do need to cost money – are profitable. And so, I think the win for the music business is to realize that music is a chance to spread an idea and the souvenir is what they need to do to make money, and my fear is that there's too much nostalgia in the music business for them to grab a hold of that quickly.

Well, I think the other thing you're going to see in the music business is extracting more revenue via licensing from the companies who value their content – webcasters and radio stations are obvious sources of extortion via licensing.

Well yes, except that if you charge too much, then people who aren't going to pay the license aren't going to spread the idea. What I'm trying to get at is, no one would ever charge Oprah to be on her show. If Oprah calls, you go on even though you're giving all this content away for free. When you sing a song on the old *Dick Clark Show*, all the people who are watching get to hear you for free. When you're on MTV, you're not charging them to put on the music video.

So what I think we're seeing is being a hit first is the goal, and it'll monetize itself if you work hard on it afterwards.

If you come to me and say, "We're not going to charge you to listen to this Rickie Lee Jones song, but we will let you join her fan club. We will tell you about her concert. We will put together all sorts of interesting opportunities for you to share this content with your friends," and all those things cost money, that makes way more sense than saying to someone, "You have to pay me money to hear this Rickie Lee Jones song."

Conclusion: Radio's Andy Grove Moment

In a classic experiment in psychology, a room of participants were handed a very simple test: A page featuring three straight lines of varying lengths. Their job was to determine which of the three lines matched a fourth line. Now this was an easy task since the lengths were so glaringly different and only one of the three lines clearly matched the length of the fourth.

But what one participant didn't know was that the other "subjects" in the room were really actors, and all of them had been instructed to give the wrong answer.

One by one they called out the answer – each one obviously wrong. The genuine participant was, of course, bewildered.

But guess what happened?

The test subjects joined the group in giving the wrong answer in at least one round 75% of the time. Time and again, they went along with the group to save themselves the embarrassment of being the odd man out.

Any group, in other words, has a tendency towards group-think. And in times of rapid transition, group-think can kill. Groups tend to lock in on one direction and doggedly cling to it no matter what. It takes openness and courage and contrarian thinking to break this stranglehold. It takes not only the necessity to change but the sheer will, too.

Former Intel CEO Andy Grove tells the story of how he decided to take his company out of the memory chip business, which was increasingly crowded by an onslaught of high-quality, low-priced Japanese chips.

But memory chips were Intel's core business. As Grove told the story in his book *Only the Paranoid Survive,* "Our priorities were formed by our identity; after all, memories *were* us." But as competition proliferated and the ink turned red, Intel needed to do something dramatic.

Grove continued, "I was in my office with Intel's chairman and CEO, Gordon Moore, and we were discussing our quandary. Our mood was downbeat. I looked out the window at the Ferris wheel of the Great American amusement park in the distance, then I turned back to Gordon and I asked, 'If we got kicked out and the board brought in a new CEO, what do you think he would do?' Gordon answered without hesitation. 'He would get us out of memories.' I stared at him, numb, then said, 'Why shouldn't you and I walk out that door, come back, and do it ourselves?'"

And so Intel, the memory chip company, became Intel, the microprocessor company, and one of the greatest business success stories of our time.

The radio industry is approaching an Andy Grove moment of its own.

Recently, consultant Fred Jacobs recruited a panel of ringers and thinkers and zealots and troublemakers for a highly anticipated discussion on this question:

"If you were president of radio, what would you do?"

Now let's leave aside the obvious implication that presidents are generally elected and are thus responsible to a slew of promises and compromises. What Fred was really looking for was this: What would you do if you were in charge?

Here is my answer to that question. Imagine this as the inaugural address I'll never get a chance to make:

People of radio, our long national nightmare is over.

I would like to join with my campaign co-chairs Jessica Biel and Jessica Alba to thank my good friends Angelina Jolie and Scarlett Johansson for their tireless efforts on behalf of my campaign.

And give it up for my chief style advisor, Diddy!

Do I know how to win an election, or what? If iBiquity had been running for this office they would have sold you a lot of polling booths that no listeners wanted to vote in!

As my first action, I want to announce an 11-point plan of action designed to transform the radio of the past into the radio of the future.

1. Recruit and Nurture Talent

The advantage for radio in the years to come will have less to do with songs and more to do with what comes between them.

I know this is contrary to what PPM ratings tell you, but PPM, like diaries, measures only that part of our competitive universe situated on the radio dial. It tells you nothing about the Accuradios and Pandoras and AOL Radios. It tells you nothing about the podcast audience. It tells you nothing about the hours of listening that vanish from the radio dial because thousands of my favorite songs are in rotation on my iPod. It tells you nothing of what surprises the Internet may bring to your car any day now.

What Arbitron ratings tell you is how well or how poorly you're doing compared to the rest of a shrinking pie. And as the pie shrinks, so will the ad dollars.

Yes, more than 90% of America still listens to the radio. But in the past 10 years, Arbitron's own statistics indicate that time listening among persons 12+ is down by 10% – and among persons 18-24 it's down by 20%.

The key to recovering that audience is offering them an alternative to the music machines that are suddenly a dime a dozen.

And that takes talent. On-air talent.

As I illustrated in the introduction to this book, we can't even name one nationally known superstar radio talent under the age of 40 – besides Ryan Seacrest.

You'd better fix this. You'd better launch an all-out campaign to discover and nurture and expand talent. The Disney Channel does it. So can you.

2. You're Not in the "Radio Business" Anymore

If you think your broadcast tower is the center of gravity for your radio station, you're wrong. Unless you can dress up that tower with a sign reading "Eiffel" and sell tickets to sucker tourists, it's time to get your head on straight.

And if you view your website as an extension of your radio station's brand rather than vice versa, you're wrong again.

Consider these questions: If you were starting your station from scratch right now, what would it look like? What would your digital strategy be? How would the broadcast capacity be integrated? Who would you be competing against? What would be your revenue models? What would your priorities be?

You need to understand this critical point: Radio is not "a broadcast," it is not "an audience," and it most certainly isn't "a tower." **Radio is the web of local relationships between advertisers and consumers mediated by your company – no matter how or where you connect those relationships.**

What you have that your competitors lack is that incredible loudspeaker called the broadcast. You can move consumers – quickly. You can economically make a client's dreams come true. And you don't need to do it with spots.

On the digital front, for example, why invest all your efforts into your radio station's website? A digital strategy should not be a billboard on the information superhighway, as most radio station websites are.

Instead recognize that you are a local media business, not a local radio station. And that means you should leverage the magic power of your loudspeaker to drive consumers to any number of local web destinations, each defining a particular local interest, all of them owned by you and none of them branded with your call letters.

Imagine a world where there are no more spots and no more rankers. Be ready to win in that world and you'll surely win in this one.

3. Experiment and Create

In biology there's a concept called "adaptive space." It's that zone of possibility poked and probed by evolution as ecological roles are filled by one species and then another. It's nature's playground for novelty and it is the way the universe works.

"Life will find a way," Jeff Goldblum said in *Jurassic Park*. Life is inherently experimental, always trying something new, passing it on if it is beneficial and squashing it if it isn't.

Rigid life forms, those set in their ways or with very narrow requirements, are unlikely to weather shocks to their environment. And that's true no matter how successful these species were prior to these shocks. The most famous example of this, of course, is the fate of the dinosaurs who ruled the world until a giant asteroid struck the planet some 65 million years ago.

Those forms of life most flexible in their capacities to cope with a changing environment are the ones which survive and thrive from one age to the next. And flexibility is about experimentation and novelty.

Once upon a time there was a TV show called *The Sopranos* which was created by one guy who became famous and was executive produced and occasionally written by another guy who did not.

That other guy went on to create a show called *Mad Men* about the hustle and bustle (in more ways than one) of a New York ad agency circa 1960.

This show never really scored with audiences during its first season and only by the skin of its teeth limped to a renewal for season two.

The folks at AMC TV who green-lit this show evidently believed in it enough to give it another chance.

Between season one and two, the critics had their say, and *Mad Men* scored two Golden Globes, including one for "Best Television Series – Drama" and a slew of other awards and nominations, including 18 Emmy noms.

Also during this time repeat airings of the first season and a DVD release sparked more interest.

Cut to: Season two.

And this headline: "Mad Men *Debut A Ratings Hit: Draws 2 Million Viewers For Highest Audience Ever.*"

That's double the first season average.

Heretofore, AMC had been best known as that cable network that runs all the old movies, but not as old as the ones on Turner Classic Movies. But in part as a result of *Mad Men*, AMC's overall ratings are up and the network is suddenly transformed into a destination, not simply a utility.

If you think about it, the similarities between a movie network that plays all familiar and popular old movies and a typical radio station running familiar and popular old songs is more than coincidental.

AMC broke from the pack by taking some tremendous risks on talent. Then AMC gave the fruit of that talent time to ripen, even when it looked as if the audience had already spoken. AMC bet on quality. They took a chance. And they put their money where their proverbial mouths were.

It seems obvious to me that the future of radio will lead us to one of two scenarios:

One: Where radio stations trim and carve their way to a bare bones utility listened to for those in the habit when other options are not available.

Two: Where radio stations bet on quality and on talent, take a chance, and put their money where their proverbial mouths are.

We have the advantage of universal access, ease of use, and strong leads in the workplace and the car in particular.

What a shame if we squander these advantages, all for lack of vision and out of an abundance of fear.

Every business – every medium – will only deteriorate over time, all other things equal. The key to long-term success and profitability is to foster new beginnings, to add new products to the industry that needs them. And adding new products requires the introduction of new ideas. And this generally means new people.

Where choice proliferates, existing choices will always erode. This is not something to be avoided or feared, it is to be embraced. Radio itself is a niche, and when old niches shrink the only solution is to build new ones – to inject new ideas into your company, your products, your team, and your strategies.

We can bemoan the lack of attention radio currently gets in the news media despite its ongoing popularity, but the news media, like consumers themselves, focus on what's new, not what's old and familiar. The more novelty – the more ideas – we pour into our brands and our strategies, the better our chances of riding a wave into the future.

Radio has an unfortunate history of penalizing those who experiment and fail. But failure and the kind of innovation these times require – the kind that will yield transformational successes – will always go hand in hand. The human family tree is cluttered with the dead branches of extinct species, but we owe our own success – our very existence – to those failures.

If our industry continues to discourage creativity and rewards only the conservative march towards yesterday, we will remain as vulnerable as dinosaurs standing idly by as a meteor crashes to the Earth.

4. HD Radio: Get Over It

It was a noble experiment – a great one, even – and it was a failure.

I was first to warn this might happen several years ago, and I discussed the numerous pitfalls facing HD radio in my previous book, *Fresh Air*.

Sadly, rather than address those many risk factors, the HD radio establishment moved instead to wage a public relations campaign urging stations to upgrade for benefits yet to be appreciated and consumers to purchase technology in which they perceived little benefit and had little interest.

Predictably, this was a recipe for failure.

The Achilles' heel of HD radio was its creation as a solution to the problems of the radio industry rather than the problems of consumers. The history of innovation proves that when you start in the wrong place, you rarely finish anywhere else.

If you were to poll the industry's rank and file today, you would discover a clear acknowledgment that HD radio is a bust, regardless of the breezy promises and proclamations of those who back this technology. In business, it's advisable to read which way the consumer wind is blowing and to avoid facing into it. And the wind is blowing at hurricane strength towards the Internet and mobile media and away from HD radio.

It's time to acknowledge the truth and move on.

The risk is that HD radio will sap our attention from the true opportunities that lay at our doorstep. The risk is that we will put our eggs in the wrong basket and end up without a basket worth owning.

5. Come to Terms with AFTRA

Shocking, but true: Most broadcasters don't understand the potentially deadly consequences of streaming their stations online.

I say "deadly" because of the way Arbitron treats the streaming version of your station versus the over-the-air version.

You probably know that your station's online stream doesn't match your over-the-air station exactly. That's because certain spots can't be run online due to the lack of a definitive compensation agreement covering streaming between the radio industry, AFTRA (the performing talent featured on the spots), and the agencies which create the spots.

Since the station and the stream don't match exactly, Arbitron rightly counts these as two different stations, since all spots must run in both places for the stations to be considered one.

And that means that every time you migrate one of your heavy listeners from your station to your online stream, you migrate their listening from the high-ranked station where every tenth of a share can determine whether or not you're in the buy, to the low-ranked station where a little more or less listening doesn't matter at all.

In other words, unless and until all spots can run in both places, your stream is likely to be cannibalizing your over-the-air station and costing you dearly in an Arbitron-ranked world.

Now sure, you could argue, the stream also acts as an introduction to your station for folks who haven't listened yet. And sure, it may be the only form of your station some listeners can or will listen to. In other words, it certainly can add new listening to your brand.

But at what cost?

Thanks to the precision of Arbitron's reporting tools, you can actually examine how many heavy users are tuned to your station online. And you can model the impact of those listeners on your overall audience share if only you could add them into your total listening audience – which you currently cannot do.

That is the *true* cost of streaming, a cost which penalizes you in ratings as your stream becomes more popular, assuming that popularity comes from listeners who would otherwise listen over-the-air.

It doesn't work this way everywhere, of course. In Canada, for example, there is no such obstacle to success. There, stations spread their streams as far and wide as they want, and every new listener counts.

This is an issue which has gotten very little visibility in the radio industry, and that has to change. Recently I discussed this topic with a major radio group head who directed me to the chairman of the radio board at the National Association of Broadcasters who, I was told, was working on a solution.

He did not return my call.

Now consider this from another point of view. This view celebrates all these unduplicated digital avails because that means we have that much more "stuff" to sell advertisers – and to sell with perfect accountability and highly detailed metrics.

The champions of this argument are creating a marketplace for digital advertising in a radio world that has never known it before, a world that must not hesitate to pursue it, no matter the cost - even if the cost is potentially a rank position in Arbitron.

I think we can have it both ways.

Having one brand with one set of spots online and off protects your Arbitron rank while not precluding you from inventing a host of digital streams with plenty of digital advertising built in. One radio station can launch as many streams as it can profitably monetize.

Now, if the spots are matched online and off, will stations bother to invest in unique digital streams filled with unique digital spots?

They'd better.

6. Prepare for Higher Royalties

"Music radio is going to be in trouble."

So said Talk Radio stalwart Sean Hannity at an event I attended not long ago. And while Sean is not known for seeing all sides of many issues with equal clarity, on this one I think he's right.

Since its very beginning, American radio has been exempt from paying royalties to song performers (though not songwriters) due largely to the tremendous promotional value of free airplay.

But as Internet radio, satellite radio, and MP3 distribution transform the music landscape – and pay performers in the process – is this exemption still valid?

Labels, which can no longer efficiently and effectively chase down their song-thieving customers, have no choice but to extract more value for their content from the licensees who have a strong incentive to pay for it.

Broadcasters, meanwhile, correctly argue that free airplay is worth something significant, thus explaining why the labels lobby so hard to get it.

Nevertheless, the sympathies in Congress and in the public arena are likely to lean towards the lowly performers and away from the corporate broadcasting giants and their agents, despite the fact that radio moves more music in one hour than all the radio alternatives are likely to move (without radio) in one year.

Radio lacks symbols as powerful as the archetypical "starving artists" the music business trots out to obscure their own megalomaniacal (or are they survivalist?) aspirations. And this will cost us.

This means radio will probably be paying more for music licensing in the future than it has in the past.

And here's exactly how we should respond:

It's time for the radio industry to levy a fee on labels equal to and opposite whatever fee is levied on radio.

It's time for *legal payola.*

There is no law against taking money from labels for airplay as long as that exchange is open.

Welcome radio's newest major-league sponsor: The music industry!

So, Mr. Music Label, we'll be happy to pay for performance – as long as you're happy to pay for play.

This argument has been sponsored by Sony Music.

7. Worship Effectiveness, Not Reach

Thanks to PPM measurement, radio reaches a lot more folks than we ever expected. Some audiences are two and three times larger than they were under diary measurement.

"Our reach approaches the reach of TV!" we proclaim, even as TV struggles to remain relevant in a splintering media world where provable impact, not reach, is the metric that increasingly attracts advertising dollars.

What does it mean to be a "reach medium," anyway?

It means you can talk to an awful lot of people by advertising on radio, more than ever before.

Unfortunately you will reach them for fewer hours per week than you ever could before, too.

The essential point that broadcasters are missing is this: All other things equal, a dramatically larger audience is a dramatically less involved one.

And in a world where you can measure the clicks and conversions of exactly the right people and exactly the right offer, where do you expect the money will move – to the larger and increasingly dispassionate and passive audience or to exactly the right people for exactly the right offer?

We're so busy trying to frame the story that Arbitron naturally gives us, we're not trying to create a story that's more compelling and in tune with the times.

Radio's challenge can be summed up in these three words:

"Make it clickable."

Text messaging is one way to do this today. Interested in an advertiser's message? Just use the SMS feature on your mobile phone for more information. I am stunned that more stations aren't investing in this. It's not right for every advertiser, of course. But we should be working harder to make it right.

The future holds more dramatic changes in the user experience of radio and the user interface itself.

Take a look at the radio dial on the average clock radio. It's pathetic! Finding a station is almost impossible if two stations are close. It's analog, it's too small, it's ugly. Yuck.

What's in store, however, is a visual interface: A new face for radio.

And that face will have CD images for songs. Images that you touch to buy. Songs themselves can be sponsored by images on a touch-screen where you can click for more information and that information will be added to your personal profile, emailed to you, or sent via SMS.

You'll have the ability to hold or rewind live radio and to save content for later.

And none of this will require HD radio.

Radio will, finally, become "clickable."

And with "clickability" comes accountability and dramatically greater relevance for the advertisers of tomorrow.

8. Everything – Everyplace

Radio's future was first glimpsed on a cold night in March of 1948, but it would take 60 years before anyone would see it.

That was the night of the 20th Academy Awards, and the world of American movie-making was on the verge of radical change – thanks in part to the advent of an alternative entertainment medium called television. The Hollywood studios would try to battle this new medium, they would try to kill it. They were dependent on audiences and any substantial decrease in those audiences would mean economic disaster. (Does this sound familiar, radio broadcasters?)

Fear was in the air that night for all the studio-heads. All but one. The one with the crazy ideas.

Walt Disney would win no awards that night, but he would be the spark that ignited his industry's future and, at the same time, point the way for our own.

Previously, all studios but Disney made their money strictly from theatrical exhibition of their motion pictures. They profited by squeezing down the costs of producing films and distributing them to their own theaters. The talent was under contract and they owned the channel of distribution from head to tail. (Does this sound familiar, radio broadcasters?)

Disney had already baffled the other studio heads after successfully creating the first full-length animated feature, *Snow White and the Seven Dwarfs*, which had once been known as "Disney's folly." Not just any movie, this was the first film to gross $100 million, the first to have a soundtrack, the first to have a merchandising tie-in, the first with multiple licensable characters.

In one fell swoop, Walt Disney had done more than create a new hit, he had created a new business model.

Today, the movie business per se is a relatively unimportant part of the business of the conglomerates which own the studios. All the majors routinely lose money on theatrical release, where the massive audiences of days gone by no longer exist. According to Edward Jay Epstein, author of *The Big Picture: Money and Power in Hollywood*, the studios make the bulk of their profits from licensing their filmed entertainment for home viewing and by leveraging that content across all entertainment channels: Video, music, television, gaming, etc. Writes Epstein, "Theatrical releases now serve essentially as launching platforms for licensing rights."

Put another way, the movie studios aren't in the movie business, they're in the content licensing business.

Like the movie business in 1948, the radio industry is in the midst of dramatic change. Audiences are in motion, alternative channels for audio entertainment and information abound, and acceptance for these channels is growing. Our industry has responded by creating a new channel of distribution which we own exclusively, HD radio, but while this will be part of the tapestry of audio options in tomorrow's entertainment firmament, it is still one channel among many. You can't keep movie fans from leaving the theaters by building a new set of theaters.

So what is the answer? Here's what Uncle Walt would tell you:

A. Create Original, Magnetic, Unique Content

"Content" is a cliché in the radio business, usually representing "that which we have on the air." But this isn't what Disney meant. If radio is merely a vehicle for the product of the music labels we are only a conduit, a distribution channel, and thus every bit as vulnerable as the theater chain owners are as they enter an age where movies will be simultaneously released on the big screen, on the small one, and at a video store near you.

"Content" means stuff we create that no one else has. "Content" means viewing radio as a source for unique, compelling programming, not as corporate radio's musical iPod. "Content" is Howard Stern, it is not the new Kelly Clarkson hit. "Content" is *Loveline* or Clear Channel's *Stripped*. Content is the Ricky Gervais podcast which is now (supposedly) the most downloaded podcast ever, with millions of cumulative listeners.

B. If "Content is King," Distribution is Queen

The goal of the radio industry should be to syndicate or license its content across all available platforms. Satellite radio is about to figure this out. Look for them to behave more like a production company than a set of pipes which must be subscribed to. Listeners will buy Howard Stern, not Sirius XM. Sirius XM may think Howard's in the Sirius XM business, but in fact the exact opposite should be true. In the future, listeners will buy the content, not the channel.

They'll be more apt to listen to what's ON your station, not your station itself.

Take a page from Ricky Gervais. His last series of his podcasts was available for sale only. That is, each of at least four 30-minute episodes retailed for $1.95 at Audible.com – the entire series sold for $6.95. If just half a million of the millions of listeners who downloaded the free episodes paid for at least one Audible.com episode, Gervais and company grossed $1 million – for as little as two hours of programming.

C. Get Set for a More Costly, Risky Business

Talk to the TV networks and they'll tell you that creating content is a risky, expensive business. But a handful of hits make all the risks worthwhile. The radio industry will have to awaken to new market realities: Investment, trial, failure, success. More programs, less programming. There will be no free lunches and no shortcuts. It will not be possible to operate multi-million dollar franchises like an FCC-licensed CD player. The bar has just been raised.

D. Follow the Vision

The big wins in radio will go to those who have the vision to see the future and take risks – those who put Wall Street's money where their mouths are. Just as the golden-age movie studios couldn't cut production expenses to create their future, you can't save your way to success either. Look for the broadcasters with big ideas and follow them.

9. Know Your Audience – By Name

Radio's opportunity, says the ever-insightful Seth Godin, is to "build relationships for the long haul with people who want to hear from you [and that's an advantage every radio station has today], over every entrepreneur that doesn't."

And the way to do this, says Seth, is to know your audience as individuals. You must know their names, their addresses, their email addresses, their tastes, their interests, etc.

If you know these things, then you are uniquely qualified to build solutions that address those needs.

"But what does this have to do with radio?" you might ask.

Everything. Because radio is not your over-the-air product. **Radio is the web of local relationships between advertisers and consumers mediated by your company – no matter how or where you connect those relationships.**

And what do you call a listener whose name and address and email and interests and passions you don't know? You call her a "stranger." And you don't have relationships with strangers, do you?

Radio is obsessed with granular observation of the ratings when we should be obsessed with granular observation of our audience. Mistake one for the other at your peril.

And how does it benefit you once you know everything there is to know about your audience as individuals?

Suddenly you're able to connect exactly the right advertiser with exactly the right listener in exactly the right time and place. And you do it by leveraging your relationship with that listener and that advertiser.

This turns "ears" into "leads" – and leads are a lot more valuable than impressions ricocheting off ears like bullets off rocks.

So not only should you be building a database, but the database should be a vital and dynamic tool, not the recipient of regular and anonymous "email blasts." The database should become a community of active participants, not a list of listeners. And your goal should be to capture 100% of the audience in your community.

10. Enter the Leaders

I wouldn't have recognized him if he hadn't been just a few feet away from me.

People were coming and going, seemingly oblivious to his presence, but it was unmistakably him. Wearing that archetypical black mock turtleneck and jeans, I had to wonder if he had anything else in his closet.

I had seen him many times before, of course. Usually in front of a super-sized screen introducing this or that technological innovation which was destined to change the world.

But here, in the flesh, Steve Jobs looked just like a regular guy.

Just like you or me.

A leader isn't someone who looks like one, he or she is someone who acts like one.

And what does it mean to act like a leader?

"If your actions inspire others to dream more, learn more, do more, and become more, you are a leader."

So said America's sixth president, John Quincy Adams, nearly two centuries ago.

Leaders motivate, they don't preach. They inspire before they retire. They act, they don't simply react.

As former Chrysler chairman Lee Iacocca once wrote, "So what do we do? Anything. Something. So long as we just don't sit there. If we screw it up, start over. Try something else. If we wait until we've satisfied all the uncertainties, it may be too late."

If you wait until the fix is easy and cheap and pain-free, it will most definitely be too late.

The truth is, we need more than leaders in radio; we need heroes.

"A hero," wrote Joseph Campbell, "is someone who has given his or her life to something bigger than oneself."

Now is the time for something bigger than a cosmetic industry re-brand, bigger than cut-rate music machines, bigger than HD-this or that.

My sincerest hope is that the leaders of our industry reading these words heed the call and rise to the challenge.

11. A Message for Wall Street

Finally, this message for my friends on Wall Street.

As president of radio, I regret to inform you that our industry is going to take a step back in order to take two steps forward.

That means we will be making some big strategic investments in our future over the next two years. These are essential investments that will diminish our profitability in the short-run.

We could do better in these quarters if we simply did nothing, proclaimed progress using some artful, symbolic language, and cut everything that doesn't have legs – along with plenty of expenses whose legs come in pairs.

We know you'd like that, because our performance would be better…

…in the short run.

But as president I am hereby deciding that any business run for the next 90 days is a business which has run out of ideas and run out of any justification to exist. Indeed, we are in business not to make you short-term dollars but to build profits over the long-run, and along the way build our communities, enrich our clients, and create happy and healthy lives for thousands of employees who deserve the very best we have to give. They, too, are our stockholders, and I'm confident they will understand.

Now I'm sure you will punish our stock, as you commonly do when you hear things you wish you didn't.

And I say: Go ahead.

Because the more you drive down our stock, the more you transform it into a bargain.

And my sincerest hope is that when that bargain pays off, many a wise investor will make lots of money, *and none of those wise investors will be you.*

Thank you for your support.

Postscript

EBay was born in 1995 when a computer programmer named Pierre Omidyar couldn't register the name of his consulting company, Echo Bay Technology Group, so he shortened the domain to "eBay."

The very first item to appear on eBay was Omidyar's laser pointer – his *broken* laser pointer.

After selling the item for $14.83, Omidyar contacted the winning bidder to make sure he understood that this laser pointer was, in fact, busted.

He did. You see, he was a collector of broken laser pointers.

There's something for everyone and, thanks to the democratizing force of digital media, there's now everyone for something. Still, radio has an advantage few other media have, regardless of where those media live.

We have the biggest, most effective megaphone in town.

Use it.

To create a new future.

While you can.

The buzz always yields to what's new and sexy, to be sure.

But that's only part of the story.

You see, an iPod never soothed your fears when a tornado leveled your neighborhood. An Internet stream never volunteered its time and money for your local community. A satellite radio station never brought your favorite music artist to town. A mobile phone never tossed you a free t-shirt at a movie screening. You never called Apple to play a game or request a song or enter a contest. Nobody at last.fm ever inflamed your political passions or solved your relationship problems or helped you handle your money. Internet radio never helped you find your way home in rush hour and never helped you know what to wear to work or school. It never made you smile or cry or feel like you're part of an extended family, singing along to the same tune and laughing along to the same jokes.

The miracle of radio is not that we play the same songs our competitors do, but that we do everything else they can't.

Radio is that friend in the dark, that playground of the mind.

Close your eyes and see what you hear.

About Mark Ramsey

Mark Ramsey provides an incisive interpretation of a media brand's place in its market and its opportunities to bust out of the pack and shine, along with all the audience research and brand development you need to get there.

Ramsey also consults media brands on navigating the challenges and opportunities of these fast-changing times.

Clients have included Clear Channel Communications, CBS Radio, Greater Media, Bonneville Broadcasting, Sirius/XM Radio, and major media players like EA Sports and Apple.

If you want to put the kind of thinking you read on every page of this book to work for your brand, whether it's in radio, TV, or new media, please contact:

Mark Ramsey
+1-858-485-6372
http://www.radiointelligence.com
http://www.hear2.com
or email ramseymark@earthlink.net

Also by Mark Ramsey:

Fresh Air: Marketing Gurus on Radio

Made in the USA